Micro-Nano Surface Functionalization of Materials and Thin Films for Optical Applications

Micro-Nano Surface Functionalization of Materials and Thin Films for Optical Applications

Editors

Ramón Escobar-Galindo
Elena Guillén Rodríguez

MDPI • Basel • Beijing • Wuhan • Barcelona • Belgrade • Manchester • Tokyo • Cluj • Tianjin

Editors
Ramón Escobar-Galindo
Applied Physics I
Universidad de Sevilla
Sevilla
Spain

Elena Guillén Rodríguez
Functional Surfaces and
Nanostructures
PROFACTOR GmbH
Steyr-Gleink
Austria

Editorial Office
MDPI
St. Alban-Anlage 66
4052 Basel, Switzerland

This is a reprint of articles from the Special Issue published online in the open access journal *Coatings* (ISSN 2079-6412) (available at: www.mdpi.com/journal/coatings/special_issues/micro-nano_surface).

For citation purposes, cite each article independently as indicated on the article page online and as indicated below:

LastName, A.A.; LastName, B.B.; LastName, C.C. Article Title. *Journal Name* **Year**, *Volume Number*, Page Range.

ISBN 978-3-0365-1658-5 (Hbk)
ISBN 978-3-0365-1657-8 (PDF)

Contents

About the Editors

Ramón Escobar-Galindo

Dr. Ramón Escobar Galindo is an Associate Professor at the University of Seville in the Department of Applied Physics I. He graduated in Physical Sciences and has a Ph.D. in Materials Science. He holds a track record of 22 years of experience in the coordination of R&D projects, both in the private and public sectors. He worked for 3 years (2014-2017) as director of the Materials program in the solar thermal energy department at Abengoa Research. Dr. Escobar has participated in more than 25 R&D projects (being the principal investigator in 13 of them) including national and European (FP6, FP7, H2020) programs as well as contracts with companies. His main lines of research have focused on the development of advanced multifunctional materials in thin films for energy applications (in particular concentrated solar thermal energy), optical and biomedical applications. He belongs to the Spanish Society of Materials (SOCIEMAT) and the Royal Spanish Physical Society (RSEF).

Elena Guillén Rodríguez

Dr. Elena Guillén Rodríguez is a materials scientist experienced in both industry and academia. She holds a Master in Interface Science and Technology and a Ph.D. in Physical Chemistry. She has vast experience in the development of the new generation of photovoltaic solar cells based on organic dyes. After her postdoc, she worked for several years for the multinational Abengoa S.L., an international leader in solar thermal plants. She worked as a senior researcher in R&D projects concerned with optical and anticorrosion coatings for high-temperature applications, new mirror designs for thermosolar applications and the new generation of perovskite solar cells. She joined the Austrian applied research company Profactor in 2017, where she is responsible for the pre-treatment technologies and is involved in Additive Manufacturing projects. Currently, she is the coordinator of two national-funded research projects and one H2020 on Additive Manufacturing for the biomedical sector.

Preface to "Micro-Nano Surface Functionalization of Materials and Thin Films for Optical Applications"

This book contains the articles collected for the Special Issue entitled Micro-Nano Surface Functionalization of Materials and Thin Films for Optical Applications in the journal Coatings (ISSN 2079-6412). These selected articles provide a meaningful overview of recent advances and concepts beyond the state-of-the-art regarding surface functionalization of materials and deposition of thin films to be used in optical applications. The aim was to cover all relevant aspects of the topic (simulation, design, fabrication, characterization and applications) with a special emphasis on non-conventional methods for surface modification of materials, combinations of mature fabrication routes with emerging technologies (i.e., additive manufacturing) and large-area fabrication concepts to pave the way to industrial utilization of the developed materials. This overview comprises the recent work of reputed scientists from Germany, Austria, Spain and India.

New developments on the scale-up deposition of transparent conductive oxides (TCO) by magnetron sputtering are addressed in the works presented by R. Mientus et al. [1] and J. Txintxurreta et al. [2]. In [1], the effect of the reactive gases on the growth of amorphous SnO_2:Ta films was studied. These films can be used as low-temperature transparent and conductive layers to protect semiconducting photoelectrodes for water splitting and also, where appropriate, in combination with more conductive TCO films (ITO or ZnO). The low-temperature deposition of ITO films is also discussed in [2]. There, it was shown that an in-depth understanding of the deposition process is essential to growing reproducible and high (electrical and optical) quality ITO films. In particular, specific oxygen flow conditions are needed for the manufacturing of efficient transparent ITO heaters to be used in the automotive industry.

E. Guillen et al. [3] described a very versatile and non-conventional preparation of titanium oxide (TiO_x) films by Filtered Cathodic Vacuum Arc (FCVA) with different stoichiometries ($0.6 < x < 2.2$) and crystalline phases (TiO, Ti_2O_3, rutile-type TiO_2 and amorphous TiO_2). In particular, rutile-type TiO_2 films could be prepared by room-temperature and unbiased FCVA, demonstrating the great potential of this method for optical applications on heat-sensitive and non-conductive substrates.

The design of Spectrally Selective Solar Absorber Coatings based on the novel combination of computational simulation and ellipsometry measurements of the optical constants is presented in [4] by K. Niranjan et al. The obtained optical constants depict a gradation in the designed W/WAlSiN/SiON/SiO_2 multilayer stack from top anti-reflection layer to substrate due to which an enhanced solar absorption of 0.955 and a low thermal emissivity of 0.10 are achieved. Furthermore, the multilayer stack exhibits an excellent wide-angle selectivity up to 58°, which makes it a potential candidate for spectrally selective solar absorber coatings.

Finally, the design of hierarchical surface structures at different scale lengths for nanoimprinting of optical nano- and micro-structures is presented in [5] by A. R. Moharana et al. This novel process allows the creation and cost-efficient replication of highly complex optical elements, and it was demonstrated using a diffusor and a diffraction pattern. Using two subsequent nanoimprint steps, a master structure was fabricated that was further used to replicate a stamp for nanoimprinting.

We are very grateful to all the authors of the Special Issue for their high-level submissions, and we hope that the papers will be useful and of interest to the readers.

References:

[1] R. Mientus, M. Weise, S. Seeger, R. Heller and K. Ellmer. Electrical and Optical Properties of Amorphous SnO_2:Ta Films, Prepared by DC and RF Magnetron Sputtering: A Systematic Study of the Influence of the Type of the Reactive Gas. Coatings 2020, 10(3), 204; https://doi.org/10.3390/coatings10030204

[2] J. Txintxurreta, E. G-Berasategui, R. Ortiz, O. Hernández, L. Mendizábal and J. Barriga. Indium Tin Oxide Thin Film Deposition by Magnetron Sputtering at Room Temperature for the Manufacturing of Efficient Transparent Heaters. Coatings 2021, 11(1), 92; https://doi.org/10.3390/coatings11010092

[3] E. Guillén, M. Krause, I. Heras, G. Rincón-Llorente and R. Escobar-Galindo. Tailoring Crystalline Structure of Titanium Oxide Films for Optical Applications Using Non-Biased Filtered Cathodic Vacuum Arc Deposition at Room Temperature. Coatings 2021, 11(2), 233; https://doi.org/10.3390/coatings11020233

[4] K. Niranjan, P. Kondaiah, A. Biswas, V. P. Kumar, G. Srinivas and H. C. Barshilia. Spectrally Selective Solar Absorber Coating of W/WAlSiN/SiON/SiO_2 with Enhanced Absorption through Gradation of Optical Constants: Validation by Simulation. Coatings 2021, 11(3), 334; https://doi.org/10.3390/coatings11030334

[5] A. R. Moharana, H. M. Au, T. Mitteramskogler, M. J. Haslinger and M. M. Mühlberger. Multilayer Nanoimprinting to Create Hierarchical Stamp Masters for Nanoimprinting of Optical Micro- and Nanostructures. Coatings 2020, 10(3), 301; https://doi.org/10.3390/coatings10030301

Ramón Escobar-Galindo, Elena Guillén Rodríguez

Editors

Article

Electrical and Optical Properties of Amorphous SnO_2:Ta Films, Prepared by DC and RF Magnetron Sputtering: A Systematic Study of the Influence of the Type of the Reactive Gas

Rainald Mientus [1], Michael Weise [1], Stefan Seeger [1], Rene Heller [2] and Klaus Ellmer [1,*]

[1] Optotransmitter-Umweltschutz-Technologie e.V, Köpenicker Str. 325, 12555 Berlin, Germany; mientus@out-ev.de (R.M.); weise@out-ev.de (M.W.); Seeger@out-ev.de (S.S.)

[2] Helmholtz-Zentrum Dresden-Rossendorf, Institute for Ion Beam Physics and Materials Research, Bautzner Landstr. 400, 01328 Dresden, Germany; r.heller@hzdr.de

* Correspondence: ellmer@out-ev.de

Received: 15 January 2020; Accepted: 24 February 2020; Published: 26 February 2020

Abstract: By reactive magnetron sputtering from a ceramic SnO_2:Ta target onto unheated substrates, X-ray amorphous SnO:Ta films were prepared in gas mixtures of Ar/O_2(N_2O, H_2O). The process windows, where the films exhibit the lowest resistivity values, were investigated as a function of the partial pressure of the reactive gases O_2, N_2O and H_2O. We found that all three gases lead to the same minimum resistivity, while the width of the process window is broadest for the reactive gas H_2O. While the amorphous films were remarkably conductive ($\rho \approx 5 \times 10^{-3}$ Ωcm), the films crystallized by annealing at 500 °C exhibit higher resistivities due to grain boundary limited conduction. For larger film thicknesses ($d \gtrsim 150$ nm), crystallization occurs already during the deposition, caused by the substrate temperature increase due to the energy influx from the condensing film species and from the plasma (ions, electrons), leading to higher resistivities of these films. The best amorphous SnO_2:Ta films had a resistivity of lower than 4×10^{-3} Ωcm, with a carrier concentration of 1.1×10^{20} cm^{-3}, and a Hall mobility of 16 cm^2/Vs. The sheet resistance was about 400 Ω/□ for 100 nm films and 80 Ω/□ for 500 nm thick films. The average optical transmittance from 500 to 1000 nm is greater than 76% for 100 nm films, where the films, deposited with H_2O as reactive gas, exhibit even a slightly higher transmittance of 80%. These X-ray amorpous SnO_2:Ta films can be used as low-temperature prepared transparent and conductive protection layers, for instance, to protect semiconducting photoelectrodes for water splitting, and also, where appropriate, in combination with more conductive TCO films (ITO or ZnO).

Keywords: reactive magnetron sputtering; transparent conductive oxide; electronic transport; doping efficiency; tin dioxide

---

1. Introduction

Tin dioxide (SnO_2) belongs to the class of wide bandgap, oxidic semiconductors that can be, in order to achieve high conductivities, doped up to high carrier concentrations (> 10^{20} cm^{-3}). SnO_2, like indium oxide and zinc oxide, also belonging to this material class, exhibits isotropic metal 5s orbitals which form the conduction band of these compound semiconductors. The isotropy of their conduction bands, different from other semiconductors, like silicon or GaAs, is advantageous for the good transport properties (high carrier mobility) of these transparent conductive oxides even in the amorphous state [1,2].

SnO_2 is much cheaper than indium oxide, but more expensive than ZnO, the other two widely used transparent conductive oxides (TCO) [3]. An inherent advantage of SnO_2 is its high chemical stability [4], making SnO_2 suitable for applications in harsh environments, for instance as a transparent and conductive electrode on photoelectrodes for water splitting or as a selectively solar transmitting coating for high-temperature solar thermal applications [5,6].

Highly conductive SnO_2 films can be prepared using different deposition methods like spray pyrolysis [7–9], evaporation [10], ion beam sputtering [11], cathode sputtering [12], reactive magnetron sputtering [13], or pulsed laser deposition [14]. Interestingly, SnO_2 films, deposited by reactive magnetron sputtering (RMS), exhibit resistivities only in the order of 10^{-3} Ωcm [15,16]. The reason for the only moderate electronic SnO_2 film quality when magnetron sputtered, is not clear. Welzel and Ellmer discussed the role of negative oxygen ions in the creation of defects in the growing films. These O^- ions are generated at the negatively charged target surface and accelerated up to high energies (some hundred eV) in the cathode sheath [17]. These high-energetic ions impinge onto the growing film and can create defects, especially oxygen interstitials (O_i), that are detrimental to the electronic film quality [18].

The band structure of SnO_2 was calculated recently by Schleife et al. [19]. They obtain a direct band gap energy of 3.65 eV, which is in very good agreement with the experimental value E_g = 3.59 eV. The conduction band of SnO_2 is derived from Sn 5s states, i.e., the CB is isotropic with a small effective electron mass of about 0.25 m_e, which is advantageous for good electron transport in SnO_2.

Due to the formation of oxygen vacancies and/or tin interstitials, which act as donors, unintentionally doped SnO_2 shows n-type conductivity. DFT calculations prove that oxygen vacancies in SnO_2 lead to shallow donor states [20].

By adjusting the oxygen partial pressure during the deposition, resistivity values as low as some 10^{-3} Ωcm were achieved by reactive magnetron sputtering at low substrate temperatures. However, these resistivity values are not stable under normal environmental conditions or when annealed in oxygen-containing atmospheres, caused by the reoxidation of the intrinsic defects, reducing the carrier concentration. Thus, in order to achieve high conductivities, SnO_2 has to be doped like the other TCO materials (see, for instance, [21–24]). The doping can be done by replacing cations (Sn) by group V elements (P, As, Sb, Bi) or by replacing anions (O) by group VII elements (F, Cl, Br, I). Mostly, antimony (Sb) and chlorine (Cl) or fluorine (F) were used for the preparation of highly conductive SnO_2 films [25–28]. The SnO_2:Cl(F) films, especially, were deposited by atmospheric pressure chemical vapour deposition (APCVD), also named spray pyrolysis, at quite high temperatures in the range of 400 to 600 °C.

SnO_2, deposited at such high temperatures, exhibit good electronic and optical properties, however, such high temperatures cannot be used for all applications, for instance, for films on temperature-sensitive substrates (plastic), or in case of the deposition of SnO_2 films onto active devices, like thin film solar cells or photoelectrodes. In these cases, low-temperature deposition processes are required, for instance reactive magnetron sputtering [29,30].

Recently, transition metals (V, Nb, Ta) have been investigated as dopants in SnO_2. Tantalum (Ta) as a dopant in SnO_2 was investigated by Kim et al., who deposited epitaxial SnO_2:Ta films on sapphire (Al_2O_3) by metalorganic chemical vapor deposition (MOCVD) at substrate temperatures between 400 and 600 °C [31]. They reported a minimum resistivity of about 3×10^{-4} Ωcm at a Ta content of 1.5 at%. Toyosaki et al. also prepared epitaxial SnO_2:Ta films by pulsed laser deposition (PLD) onto single-crystalline TiO_2 (rutile) substrates [32]. They reported a minimum resistivity of 1.1×10^{-4} Ωcm at a Ta content of 5 at% for the SnO_2:Ta films, deposited at 800 °C. The use of single-crystalline substrates and the high growth temperatures prohibit these approaches for the large-area deposition of cheap SnO_2:Ta films. Nakao et al. tried to overcome some of these limitations by depositing SnO_2:Ta films onto glass substrates which have been coated with a 10 nm-thin TiO_2 seed-layer [33]. After the deposition onto unheated substrates, these amorphous films were annealed in situ at 600 °C to achieve resistivities as low as 1.9×10^{-3} Ωcm. Weidner et al. recently compared the two dopants antimony and tantalum in SnO_2 films deposited by magnetron sputtering [34]. They achieved a significantly lower resistivity of SnO_2:Ta (5.4×10^{-4} Ωcm) compared to SnO_2:Sb films by a factor of 3.

SnO_2 films, deposited at low temperatures, are typically X-ray amorphous [35,36]. Amorphous oxide films exhibit the advantage of the absence of grain boundary effects on the electrical transport [1]. Such amorphous oxide films in the In-Ga-Zn-O system with low carrier concentrations have been used successfully for the preparation of transparent and flexible field effect transistors, where the absence of grain boundaries in the channel regions increased the field effect mobility significantly [37]. Recently, indium oxide films were investigated in detail with respect to the relation between structural and electronic transport properties [38]. Buchholz et al. analyzed the transition from amorphous to crystalline In_2O_3 films both experimentally and theoretically.

Another beneficial effect of the absence of grain boundaries in amorphous films is that amorphous films typically exhibit lower diffusion coefficients. This property of amorphous films can be exploited for use in protective films [39].

In this article, the deposition of X-ray amorphous, Ta-doped SnO_2 films prepared by reactive magnetron sputtering from a ceramic target is investigated systematically. The depositions were performed onto unheated glass substrates in order to explore the possibility of depositing these SnO_2 films onto temperature-sensitive substrates (plastics) or sensitive underlying films or devices (thin film solar cells or photoelectrodes for water splitting). We investigated in detail the effects of the plasma excitation (DC or RF, 13.56 MHz) and the type of the reactive gas (O_2, H_2O, N_2O) on the electrical properties of the SnO_2:Ta films. Due to its superior chemical stability, such SnO_2:Ta films are intended for applications as transparent and conductive protection layers [39].

2. Experimental

The SnO_2:Ta films were prepared at different substrate temperatures by reactive magnetron sputtering from a ceramic target in a Leybold Z400 sputtering system, equipped with a fast load-lock. The base pressure of the system is about 5×10^{-4} Pa. While the reactive gases O_2 and N_2O were fed into the sputtering chamber in the conventional way with mass flow controllers (MFC) from gas bottles, the water vapour (p_{H2O} = 23 hPa at 20 °C) was fed into the chamber by a special low-pressure MFC from the liquid H_2O reservoir, a quartz container held at room temperature.

The ceramic $Sn_{0.98}Ta_{0.02}O_2$ target (purity 4N, supplier: EVOCHEM Advanced Materials GmbH) with 75 mm diameter was bonded on a Cu cooling well. It was sintered under reducing conditions, thus also allowing DC plasma excitation. The depositions were done at a target-to-substrate distance of 55 mm onto stationary substrates. Before the deposition, a substrate pretreatment with an RF argon plasma was performed for 1 min (p_{Ar} = 1.8 Pa, P_{RF} = 100 W, $V_{DC} \approx 1000$ V). The substrate holder could be heated by a resistance heater up to glass substrate temperatures of about 400 °C, measured with a thermocouple.

The film thickness was measured using a DEKTAK profilometer. The optical parameters were obtained by spectroscopic ellipsometry in the photon energy range from 1.5 to 4.2 eV (SE 850, Sentech GmbH, Berlin, Germany). These spectra were fitted and analysed with the commercial optical analysis program 'SpectraRay3' from Sentech. The optical transmittance (T) and reflectance (R) spectra were measured with an UV/Vis double-beam spectrometer (Cary 05E, Varian) in the spectral range from 200 to 3200 nm (6.2 to 0.38 eV).

The structural properties of the SnO_2:Ta films were analyzed by X-ray diffraction (XRD) with a silicon-stripe detector (D2 Phaser with a Cu-anode and a Lynxeye detector, Bruker AXS) in Bragg–Brentano geometry. The diffraction peaks were fitted by double peaks with Lorentzian peak-shape, thus yielding the positions and widths of the two CuK_α peaks.

The thickness-averaged elemental composition was measured using Rutherford backscattering spectrometry (RBS) on samples deposited onto glassy carbon substrates. He ions with an energy of 1.7 MeV at normal incidence were used for the analysis. Typically, a $^4He^+$ charge of 10 μC was used to collect RBS spectra with good counting statistics. The RBS spectra were simulated with the software SIMNRA [40]. In order to determine the electron concentration and the Hall mobility, Hall and conductivity measurements were performed with a home-built setup at room temperature at a magnetic flux of 0.74 T. The 10 × 10 mm^2 samples were contacted in the corners in the van der Pauw geometry. The radial profiles of the electrical parameters were measured in order to investigate the role of negative oxygen ion bombardment which is radially inhomogeneous [41].

The cross-section of the SnO_2:Ta films was prepared with Argon ion beam polishing and investigated with transmission electron microscopy (TEM Zeiss LIBRA 200FE) in the bright-field mode including diffraction contrast and zero-energy-loss filtering.

Post-annealing treatments were carried out with an IR-lamp heater (maximum intensity at a wavelength of about 1.5 μm) in vacuum ($p = 1 \times 10^{-2}$ Pa) or in hydrogen atmosphere ($p_{H2} = 200$ hPa) at a temperature of T = 500 °C. Prior to the annealing, the vacuum chamber was evacuated to a vacuum pressure of about 1×10^{-4} Pa. The heating rate was about 2 K/s and the time at the set temperature was about 2 min.

3. Results and Discussion

3.1. Deposition Characteristics

Figure 1a shows the deposition rate as a function of the discharge power for DC and RF (13.56 MHz) plasma excitation. For both discharge modes, the rate increases linearly with the discharge power, as expected [30], and is independent of the type of the reactive gas (O_2, N_2O, H_2O). While the linear fit curves start at the origin of the x axis for DC excitation, the RF power where the deposition rate is larger than zero is shifted to a value of about 20 W_{RF}. This is caused by the fact that the discharge voltages (i.e., the target voltages) are much lower for RF plasma excitation compared to the DC discharge [30]. For RF powers lower than 20 W, the target voltage is lower than the sputtering threshold voltage [42]; thus, no deposition can take place. The deposition rates for RF sputtering are much lower compared to the DC deposition rates; the slopes of the linear rate(power) curves in Figure 1a are 0.24 and 0.66 nm/(min·W) for RF and DC excitation, respectively. This is caused by two effects:

- The lower discharge voltages leading to lower sputtering yields;
- The fact that, in the RF case, the target is only at a sufficiently high negative potential every second half-wave, allowing an acceleration of positive Ar ions to the target surface [43].

Figure 1b displays the partial pressures of the reactive gases O_2, N_2O, H_2O, where the deposited films exhibit the resistivity minimum as a function of the discharge power. It can be seen that more reactive gas is needed with increasing discharge power, i.e., increasing deposition rate. This can be explained by the fact that a certain amount of (reactive) oxygen per deposited amount of metal atoms is needed to prepare highly conductive SnO_2:Ta films. This behaviour is also known for other TCO materials, for instance, for ZnO [44]. The amount of reactive gas needed to prepare highly conductive films depends strongly on the type of reactive gas, of the mode of plasma excitation and the deposition temperature. This is due to the different amounts of oxygen, supplied by the different reactive gases. Oxygen gas itself contains the highest relative amount of oxygen compared to N_2O and H_2O; thus, the lowest partial pressure is needed for pure O_2. Although the molecules N_2O and H_2O contain the same relative amount of oxygen (one third), there is still a difference for both reactive gases which can be explained by the different chemical action of nitrogen (N_2O) and hydrogen (H_2O): while nitrogen is almost nonreactive, hydrogen is reducing the growing film. Therefore, in case of H_2O as reactive gas, a higher amount of oxygen is needed in order to have the same oxidizing effect as N_2O. One also has to take into account that the dissociation energies of the stable molecules O_2 and H_2O (≈5.1 eV [45]) are much higher in comparison to the metastable N_2O molecule ($E_{dis} \approx 1.6$ eV [46]) which shifts the resistivity minimum for N_2O nearer to that of O_2.

This is clearly visible if the necessary reactive gas partial pressure is plotted versus the deposition rate, as shown in Figure 1c. Besides the effects of the reactivity and composition of the reactive gas, the effect of the different degrees of gas dissociation in a DC and an RF discharge can also be derived from Figure 1c. For the case of DC plasma excitation, the films were deposited both at room temperature as well as at a substate temperature of 400 °C. Obviously, the reactivity of the gas (O_2 in this case) is significantly improved at a higher substrate temperature, thus a lower oxygen partial pressure is required at a higher deposition temperature.

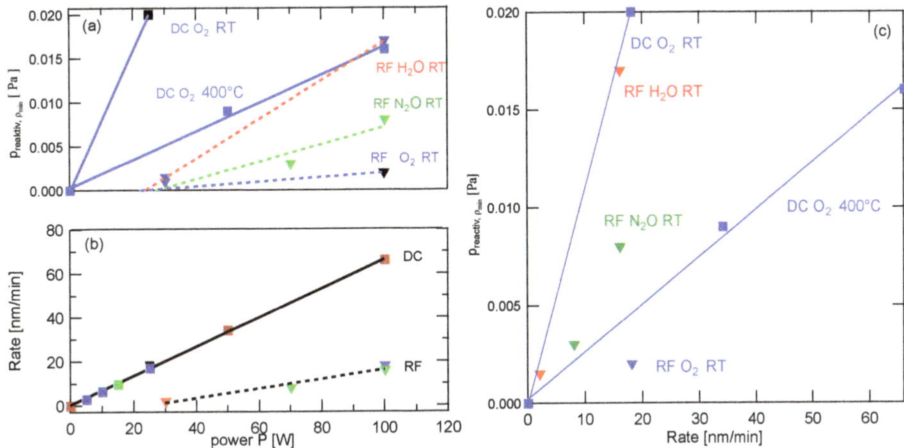

Figure 1. (**a**) Partial pressures of the reactive gases where the resistivity minimum was found as a function of the discharge power. (**b**) Deposition rate as a function of the discharge power for DC and RF (13.56 MHz) plasma excitation. Reactive gases used were O_2, H_2O and N_2O. (**c**) Partial pressures of the reactive gases where the resistivity minimum was found as a function of the deposition rate. Deposition parameters: p_{total} = 0.5 Pa; square/ triangular symbols: DC/RF excitation; reactive gases: O_2 (blue), N_2O (green), H_2O (red); black fit lines: room temperature deposition (RT), red fit lines: 400 °C deposition.

3.2. Phase Composition

The XRD patterns of the as-deposited films, displayed in Figure 2, exhibit no diffraction peaks; this means that these films are X-ray amorphous at film thicknesses below about 200 nm. While thicker films (≈500 nm) deposited by DC magnetron sputtering are still X-ray amorphous, the thicker films, prepared by RF magntron sputtering, exhibit X-ray peaks, corresponding to the tetragonal (rutile-like) structure of cassiterite (SnO_2) (powder diffraction file PDF®_01_070_4175, The International Centre for Diffraction Data, Newton Square, USA).

While the X-ray amorphous DC-sputtered films had a constant resistivity of about 5×10^{-3} Ωcm, the films which started to crystallize at larger thicknesses showed an increased resistivity for a thickness of about 500 nm (see below). The Raman spectra (not displayed here) of as-deposited SnO_2:Ta films show only weak, broad features ascribed to the glass substrates. Since Raman spectroscopy is able to detect structural features/phases for crystallite sizes in the few nm range, i.e., smaller than is possible by XRD, we conclude that the SnO_2:Ta films are really amorphous.

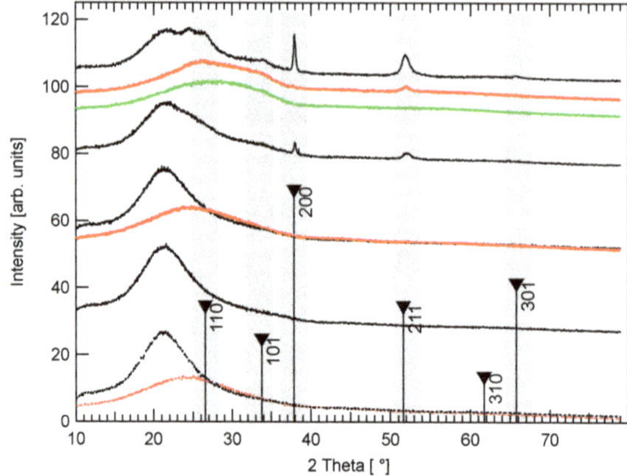

Figure 2. XRD patterns of SnO_2:Ta films deposited by RF (thin, black lines) and DC (thick, red and green lines) magnetron sputtering in Ar+O_2 atmosphere for different durations. Deposition parameters: RF: P = 100 W, p = 0.5 Pa, t = 2, 5, 13 and 23 min (from the bottom); DC: P = 25 W, p = 0.5 Pa, t = 5 and 28 min. Film thicknesses increase from about 40 to 450 (500, DC) nm with increasing deposition time. Thick, DC-sputtered films were deposited without (red curve) and with (green curve) three pauses. For comparison, the XRD patterns for the bare substrates quartz (Suprasil, black) and borosilicate glass (D263T, red) are plotted as dotted lines at the bottom. The bars at the bottom of the figure display the diffraction pattern according to the powder diffraction file PDF®_01_070_4175 for SnO_2-cassiterite (The International Centre for Diffraction Data, Newton Square, USA).

3.3. Resistivity

The reactive gas partial pressure dependence of the resistivity, the carrier concentration and the Hall mobility of the SnO_2:Ta films are displayed in Figure 3. While the discharge voltage and the deposition rate are nearly constant in the investigated partial pressure range, the resistivity exhibits a strong decrease with increasing oxygen partial pressure. A resistivity minimum is found at an oxygen partial pressure of about 2×10^{-2} Pa; further increasing the O_2 partial pressure leads to a slight increase in the resistivity.

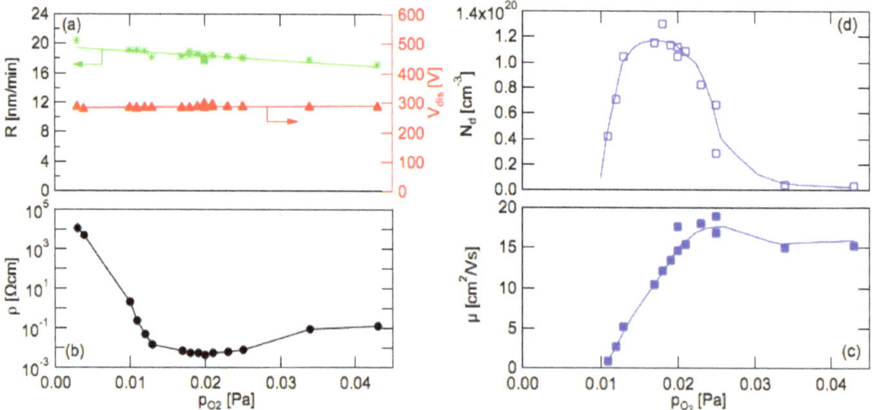

Figure 3. (**a**) Deposition rate (green) and discharge voltage (red), (**b**) resistivity (**c**) mobility and (**d**) carrier concentration as a function of the reactive gas partial pressure. Deposition parameters: P_{DC} = 25 W, p_{total} = 0.5 Pa, reactive gas: O_2, RT deposition. The lines in (**a**) are linear fit curves, while the lines in (**b**–**d**) are thought of as guides to the eye.

The resistivity minimum comes about by a local maximum of the carrier concentration and a maximum Hall mobility at the oxygen partial pressure where the minimum resistivity is obtained.

The effects of the different plasma excitation modes (DC or RF, 13.56 MHz) on film resistivity are shown in Figure 4a. It can be seen that at the same deposition rate, realized by different discharge powers for DC (25 W) and RF excitation (100 W), the O_2-partial pressure where the minimum resistivity is obtained is significantly lower for RF-, compared to DC-excitation. This, we ascribe to the much higher density of reactive oxygen species (O, O*, O^+) in the RF discharge, which was measured already in 1998 by us for the reactive magnetron sputter deposition of ZnO:Al films [47]. The minimum resistivity is about the same for both excitation modes ($\rho_{min} \approx 5 \times 10^{-3}$ Ωcm); however, the width of the minimum is significantly wider in case of DC plasma excitation.

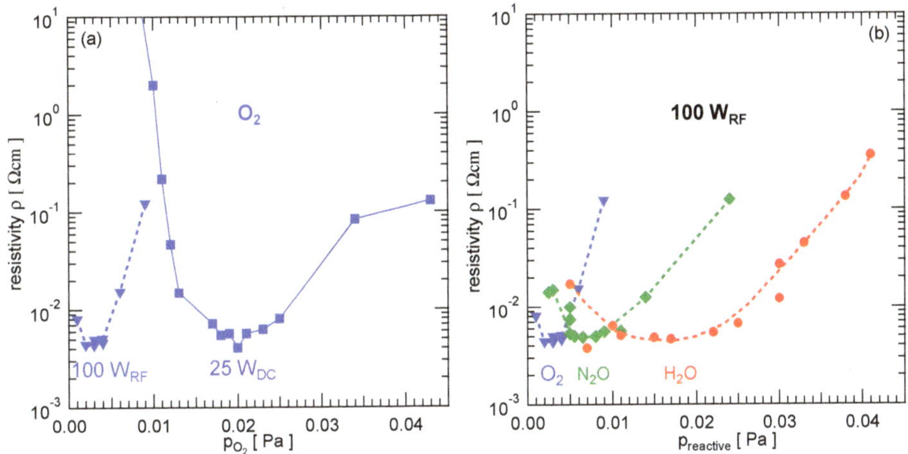

Figure 4. (**a**) Partial pressure dependence of the resistivity for the reactive gas O_2 for DC and RF plasma excitation. The deposition rate is the same in both cases (R ≈ 18 nm/s). (**b**) Dependence of the resistivity on the partial pressure of the reactive gases O_2, N_2O and H_2O for RF plasma excitation. Deposition parameters: P_{DC} = 25 W, P_{RF} = 100 W, p_{total} = 0.5 Pa, RT deposition.

Figure 4b shows the reactive gas partial pressure dependence of the resistivity for the three reactive (oxygen-containing) gases O_2, N_2O and H_2O. For all three gases, a resistivity minimum exists with nearly the same value. The position and the width of the $\rho(p_{reactive})$ curves are clearly different for the three oxygen-containing reactive gases. This is plausible, taking into account the different molar concentrations of oxygen in the different molecules—see also Section 3.1. Using H_2O as a reactive gas, the resistivity minimum occurs at the highest partial pressure, although H_2O contains the same relative amount of oxygen as N_2O. This can be explained by the reductive effect of hydrogen, thus requiring a higher partial pressure to have the same oxidizing effect, as in the case of N_2O. An advantage of H_2O as reactive gas is the wider resistivity minimum, making it easier to adjust the reactive gas partial pressure.

In order to analyze the carrier transport process, the Hall mobility is displayed as a function of the carrier concentration, see Figure 5a–c. It can be seen that, independently of the used reactive gases, the mobility first increases steeply with increasing reactive gas partial pressure, reaching maximum values of about 15–20 cm²/Vs at carrier concentrations of about 1×10^{20} cm^{-3}. The strong increase in the electron mobility with increasing reactive gas partial pressure at a nearly constant carrier concentration (see the arrows in Figure 5) can be tentatively explained by a higher defect density (Sn or SnO inclusions) at low reactive gas partial pressure. These defects act as scattering centers, thus reducing the carrier mobility. A theory for the explanation of this steep μ(N) dependence has yet to be developed.

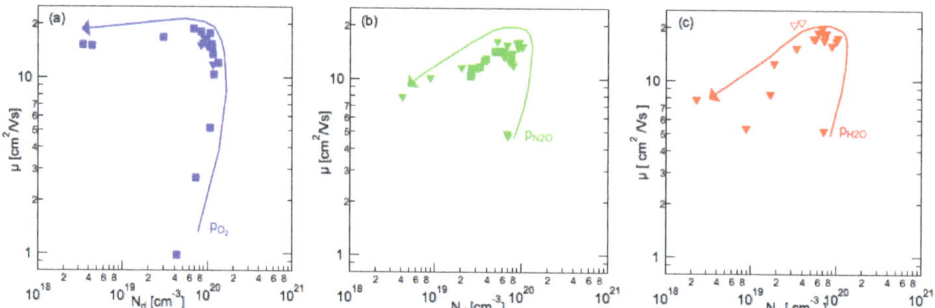

Figure 5. Dependence of the Hall electron mobility on the carrier concentration for SnO_2:Ta films sputtered with different reactive gases: **(a)** O_2 (blue) **(b)** N_2O (green), and **(c)** H_2O (red). The lines with arrows in the figures indicate the increasing reactive gas partial pressure. Deposition parameters: ■ 15 W_{DC}, ■ 25 W_{DC}, ▼▼▼ 100 W_{RF}, ▽ 30 W_{RF}, p_{total} = 0.5 Pa, RT.

Further increasing $p_{reactive}$ leads to a gradually decreasing mobility, less for O_2, and stronger for N_2O and H_2O. The carrier concentration decreases much more, by nearly two orders of magnitude. This behaviour is similar to that of other amorphous semiconducting oxide films, for instance, $InGaZnO_4$ [37,48]. According to Nomura et al., the electronic transport in these amorphous oxides is governed by a percolation conduction over a distribution of potential barriers at the conduction band edge. Other than in single- or polycrystalline semiconductors, the band edges are no longer straight over larger volumes of material, i.e., the amorphous oxides exhibit small potential barriers on the nanoscale, caused by the randomness of the amorphous material in general and by the random distribution of defects (Ta_{Sn}, O_i, V_O etc.) especially.

The maximum carrier concentration is relatively low, compared to other TCO materials (ITO, ZnO). Taking into account the concentration of Ta in the SnO_2:Ta films (Ta/Sn ≈ 0.5 at%), the electrical activation of Ta, assuming that Ta on tin lattice sites (Ta_{Sn}) acts as a donor, is quite low, in the order of 22%, using the molecular number density of SnO_2 of 3.125×10^{22} molecules/cm^3 and the maximum carrier concentration of about 1.2×10^{20} cm^{-3}. The quite low carrier concentrations of the SnO_2:Ta films could be caused by an inactivation/compensation of the donors (Ta_{Sn}) by other defects, already mentioned above (O_i, V_O etc.). Similar compensation effects were recently reported by T-Thienprasert et al. and by us for Al-doped ZnO [49,50].

3.4. Film Composition (RBS)

The film composition was determined by Rutherford back-scattering analysis (RBS). To this purpose, the SnO_2:Ta films were deposited onto glassy carbon (Sigradur) substrates which produce, due to the low atom mass of carbon, a very low background for the backscattering peaks of oxygen, tin and tantalum, thus allowing an exact determination of the average film composition [51].

Typical RBS spectra of two SnO_2:Ta films on logarithmic scale are shown in Figure 6a. Besides the edge of the carbon substrate, the peaks for O, Sn and the shoulder for the dopant Ta are clearly visible. At channel numbers around 600, a small peak is visible that belongs to argon. It is caused by two effects:

(i) Ar is implanted into the glassy carbon during the surface cleaning of the substrates by the Ar-RF plasma treatment (narrow peak at channel 590);
(ii) Ar is included (buried) into the growing film during the deposition of the SnO_2:Ta film. This is caused both by the continuous coverage of the film surface by neutral argon atoms from the sputtering atmosphere as well as the bombardment of the film surface by energetic Ar ions, atoms and metastable atoms [52,53] which are then covered by the film atoms.

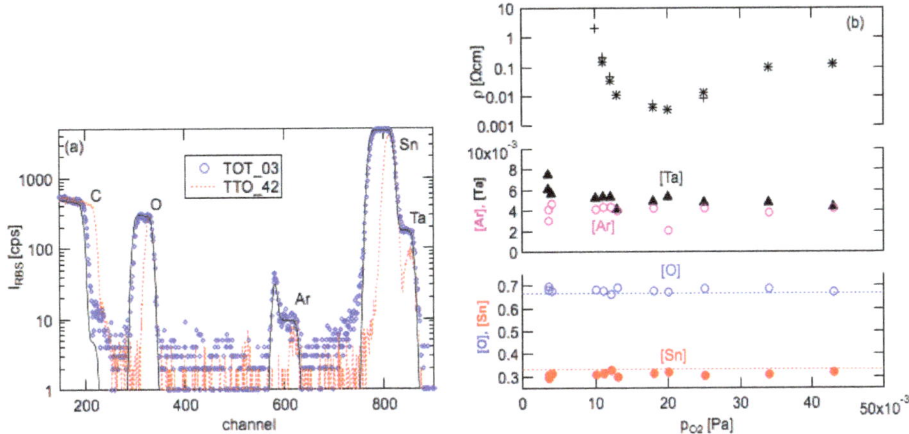

Figure 6. (**a**) Rutherford backscattering spectra of two SnO_2:Ta films on a logarithmic intensity scale displaying the backscattering peaks/shoulders for the elements Ta, Sn, Ar, O and C (with decreasing energy, i.e., channel number). The two films were deposited at low- (TOT_03) and high- (TTO_42) oxygen partial pressures. The thin line is the simulated Rutherford backscattering spectrometry (RBS) curve of the film TOT_03. (**b**) Average chemical composition as a function of the reactive gas pressure in comparison to the resistivity of the films.

The concentration of argon in the films is about 0.4 at% for films deposited at room temperature. For the SnO_2:Ta films deposited at a high temperature (400 °C), the Ar content is much lower (≈ 0.1 at%), caused by the higher desorption of the Ar atoms during the film growth [54,55].

By integrating the peaks for O, Sn and Ta, the average composition of the films was determined, displayed in Figure 6b as a function of the oxygen partial pressure. It is clearly visible that the dopant (Ta) concentration does not depend on the reactive gas partial pressure and the type of gas. It is nearly constant at a concentration of 1.5 at%. This means that the observed variation in the electrical parameters as a function of the reactive gas pressure is not caused by a variation in the chemical dopant amount, but by a varying electrical activation of the dopant Ta and/or by compensation of the electrons by acceptor-like defects—see, for instance, [49,50].

The oxygen-to-metal ratio is almost constant within a measurement accuracy of about ≈2 at%. Since we observe a strong variation in the electrical parameters when varying the reactive gas partial pressure, this has to be caused by an oxygen concentration variation which cannot be detected by RBS.

3.5. Thickness Dependence

Most of the deposited SnO_2:Ta films had thicknesses of about 100 nm. For applications where low sheet resistances are needed, thicker films have to be deposited. Therefore, the thickness dependence of the electrical film properties was investigated. Figure 7a shows the dependence of deposition rate, resistivity, carrier concentration and Hall mobility on the film thickness up to about 450 nm. As expected, the deposition rate is constant. In contrast to what is known for other TCO materials (see Figure 7b, [56]), the resistivity increases significantly with increasing film thickness, caused mainly by the decrease in the carrier concentration. A similar behaviour was observed by Minami et al. and by Brousseau et al. for undoped SnO_2 films, also deposited by RF magnetron sputtering [57,58].

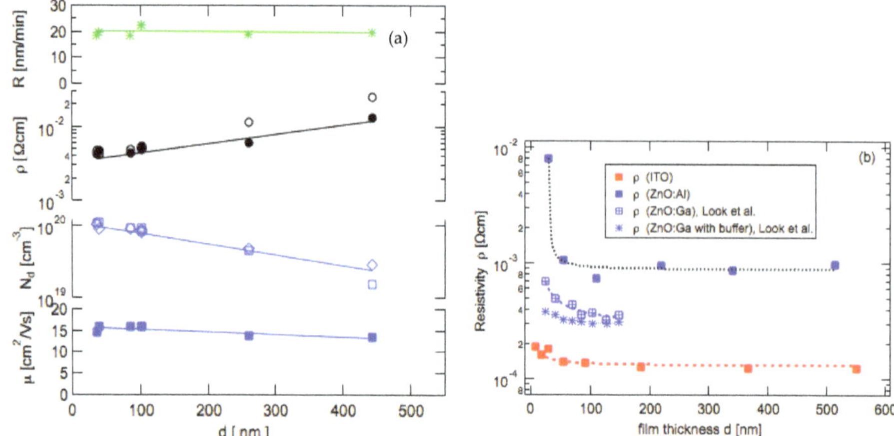

Figure 7. (**a**) Thickness dependence of the deposition rate and the electrical parameters of SnO_2:Ta films. Deposition parameters: P_{RF} = 100W, p = 0.5Pa, RT, O_2. (**b**) Thickness dependence of the resistivity of ZnO:Al and ITO films, taken from reference [56].

The well-known behaviour of a decreasing resistivity with increasing film thickness can be explained by an increasing grain size and the decreasing effect of surface and interface carrier scattering, described by the Fuchs–Sondheimer theory, see [56,59,60].

To explain the contrary behaviour of our SnO_2:Ta films, one has to take into account that the films with larger film thicknesses (> 200 nm) are no longer X-ray amorphous, i.e., these films are, at least partly, polycrystalline with grain boundaries.

Figure 8 shows a TEM cross-sectional picture of a 500 nm thick SnO_2:Ta film, deposited by RF magnetron sputtering. It can clearly be seen that the film is composed of an amorphous part, directly grown on the glass substrate. After a certain thickness is reached, varying locally, a spontaneous crystallization sets in and the SnO_2:Ta growth is transformed to polycrystalline growth. A similar behaviour was observed by us earlier for the growth of indium–tin oxide (ITO) films by reactive magnetron sputtering from a metallic InSn10wt% target [61]. In this case, the films were deposited onto intentionally unheated substrates and the increase in the substrate temperature occurred during the film growth.

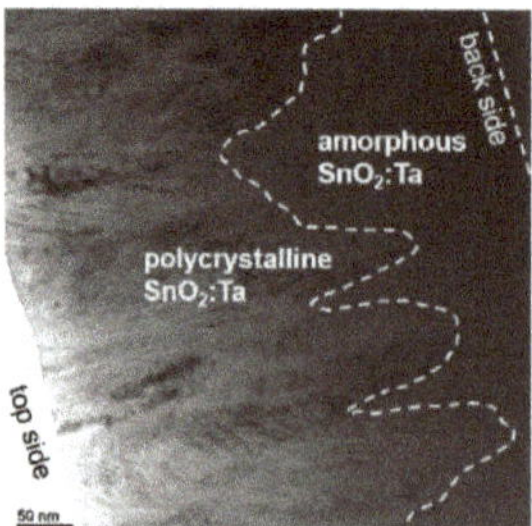

Figure 8. Cross-sectional transmission electron micrograph of a ≈ 500 nm thick SnO_2:Ta film. The flat interface to the glass substrate and the corrugated interface between the amorphous (bottom) and the polycrystalline (top) parts of the film are marked by dashed lines. The image was taken in the bright field modus, including diffraction contrast and zero-energy-loss filtering. Deposition parameters: P_{RF} = 100W, p = 0.5Pa, RT, Ar+N_2O.

The crystallization of the thicker SnO_2:Ta films is due to the increasing substrate temperature during the deposition, caused by the total energy flux to the growing film [62,63]. Based on the measured energy flux, it is estimated that the substrate temperature reaches values > 100 °C. That the increased substrate temperature causes the film crystallization was proven by depositing some films stepwise, making a pause for cooling down every 100 nm. Thick films (500 nm) deposited in this way were X-ray amophous and had resistivity values comparable to thin films.

3.6. Radial Profiles

The radial profiles of the electrical film properties were analyzed for DC and RF plasma excitation. Such profiles are an important aspect for magnetron sputtering deposition onto a stationary substrate, since the plasma/ion assistance of the film growth varies radially significantly, caused by the plasma torus which is formed in front of the target [18]. For the deposition of TCO films it is often reported that radial profiles of the resistivity exhibit pronounced maxima or minima opposite the erosion groove of the target which are formed due to the torus-like shape of the magnetron plasma—see, for instance, [18,64–66].

Figure 9 displays the radial distributions of the electrical parameters and the film thickness for DC and RF plasma excitation. While the radial thickness distributions for both excitation modes are almost the same, exhibiting the expected bell-shaped curve, the resistivity distributions in the as-deposited state are clearly different for DC and RF plasma excitation. While the resistivity of the RF-deposited films is only slightly varying over the diameter of the substrate, the resistivity distribution of the DC-deposited films shows a significant increase in ρ towards the border of the substrate by a factor of about 3. The regions, where the resistivity is high, coincide with the radial position of the erosion groove of the target. This points to the effect of a high-energetic negative ion bombardment of the growing film, as discussed recently by us [17,18].

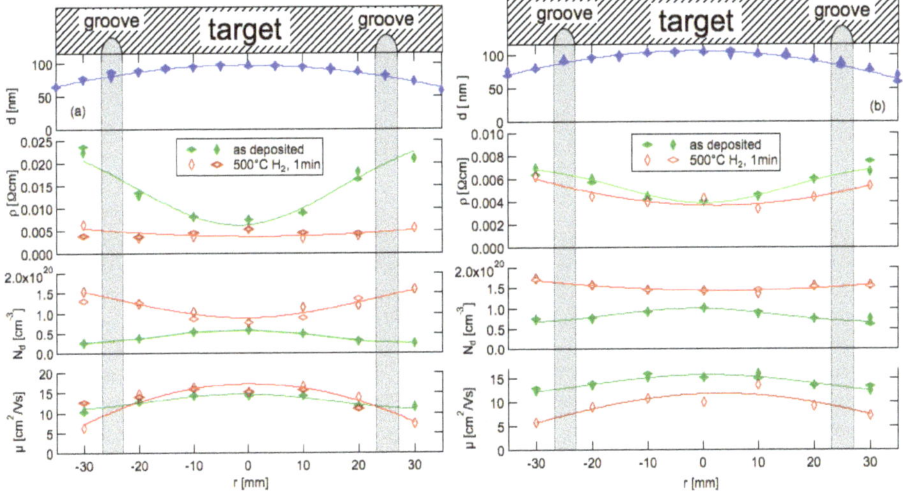

Figure 9. Radial profiles of film thickness, resitivity, carrier concentration and Hall mobility for (**a**) DC plasma excitation (P_{DC} = 15 W, V_T = 317V, 0.5Pa, RT, N_2O, R = 10 nm/min) and (**b**) RF plasma excitation (P_{RF} = 100 W, V_T = 92 V, 0.5Pa, RT, N_2O, R = 17 nm/min.). Vertically and horizontally elongated diamonds mark datapoints measured over two diameters in the substrate plane, perpendicular to each other. The red symbols and curves display the electrical parameters after annealing at 500 °C in H_2 for 1 min. The lines are Gaussian fits to the datapoints.

It has become known in recent years that, in sputtering processes with electronegative elements (for instance O, F, Cl, S, Se, but also Au, Pt, Ag, Sn, Bi—see reference [67] for the electron affinity values) negative ions are accelerated away from the negatively biased target surface towards the substrate, i.e., the growing film. Due to the high discharge/target voltages, even in the modern magnetron sputtering systems, the energies of these negative ions are in the range of some 100 eV, which therefore leads to the generation of defects in the growing film. In the beginning of the sputtering technology, before the invention of the magnetron sputtering sources, the discharge voltages were much higher (some kV) which led to a very high-energy bombardment of the growing films, sometimes ending up not in film deposition but in film and substrate etching [68]. Short-term post-anneling of the SnO_2:Ta films at 500 °C in the hydrogen atmosphere of the dc-sputtered films reduces their resistivities even in the region opposite the target groove, while the resistivity of rf-sputtered films was nearly unaffected.

3.7. Transmittance and Reflectance

The transmittance and reflectance spectra of SnO_2:Ta films are shown exemplarily in Figure 10a,b. With increasing N_2O partial pressure, the amount of non-oxidized Sn in the films is reduced and the films become more transparent in the visible spectral range. The absorption edge shifts to smaller wavelengths. This means that the optical band gap energy is increased, caused by the Burstein–Moss effect [69,70].

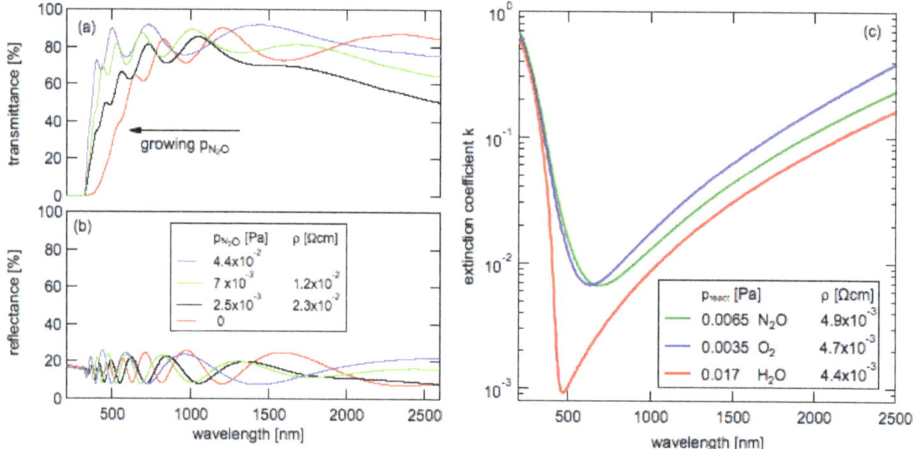

Figure 10. (**a**) Transmittance T and (**b**) reflectance R spectra in the wavelength range from 200 to 2600 nm of SnO_2:Ta films, deposited at different N_2O partial pressures. (**c**) Spectral extinction coefficients of 3 SnO_2:Ta films, deposited with different reactive gases and nearly the same minimum resistivity of about 4.7×10^{-3} Ωcm. Deposition parameters: P_{RF} = 100 W, 0.5Pa, RT.

From the transmittance and reflectance curves (see Figure 10a,b), the optical constants n and k have been calculated. The extinction coefficient k of three SnO_2:Ta films is shown in Figure 10c as a function of the wavelength. These films were prepared with the three reactive gases O, N_2O and H_2O under conditions that yielded minimum resistivities of about 4.7×10^{-3} Ωcm. While the films deposited with O or N_2O exhibit a minimum k value of about 7×10^{-3} in the visible spectral range, the films deposited in Ar+H_2O show a much lower extinction coefficient, lower than 1×10^{-3}. A tentative explanation of this striking effect is the passivation of deep defects in the band gap of the SnO_2:Ta films by the hydrogen from the sputtering atmosphere when using H_2O as a reactive sputtering gas.

4. Conclusions

Conductive and transparent SnO_2:Ta films were prepared at low substrate temperatures (<100 °C) by reactive magnetron sputtering from a ceramic target in various gas mixtures: Ar/O_2(N_2O, H_2O). The width of the process window with respect to the reactive gas partial pressure depends on the type of the reactive gas; it is wider for N_2O and H_2O, mostly due to the dilution of the oxygen content in the compound gases. The SnO_2:Ta films are X-ray amorphous in the as-deposited state. By heating the films to temperatures above 500 °C, the films start to crystallize, accompanied by an increased resistivity. For RF deposition (larger energy flux to the growing film) at larger film thicknesses (d >≈ 150 nm), i.e., longer deposition times, crystallization occurs during the deposition, caused by the substrate temperature increase due to the energy influx from the condensing film species and the plasma. While the amorphous films are remarkably conductive ($\rho \approx 5 \times 10^{-3}$ Ωcm), the crystallized films exhibit higher resistivities due to grain-boundary-limited conduction. The best amorphous SnO_2:Ta films had a resistivity of better than 4×10^{-3} Ωcm with a carrier concentration of 1.1×10^{20} cm^{-3}, and a Hall mobility of 16 cm^2/Vs. The sheet resistance was about 400 Ω/□ for 100 nm films and 80 Ω/□ for 500 nm thick films. The average optical transmittance from 500 to 1000 nm is greater than 76% for 100 nm films, where the films, deposited with H_2O as a reactive gas, exhibit a slightly higher transmittance of 80%.

The crystallization of the SnO_2:Ta films is not only detrimental for the electrical transport properties of the films; grain boundaries are also diffusion paths for atoms and molecules in gaseous or liquid environments. Since these films were deposited at low temperatures, they are amorphous, thus improving the resistivity against degradation and etching.

These X-ray amorpous SnO_2:Ta films can be used as low-temperature transparent and conductive protection layers, for instance to protect semiconducting photoelectrodes for water splitting, and also, where appropriate, in combination with more conductive TCO films (ITO or ZnO).

Author Contributions: Conceptualization, R.M., S.S. and K.E.; methodology, R.M., S.S., R.H. and M.W.; formal analysis, R.M., S.S., R.H. and K.E.; investigation, R.M. and M.W.; data curation, R.M., M.W. and K.E.; writing—original draft preparation, K.E. writing—review and editing, R.M., S.S., M.W. and K.E. All authors have read and agreed to the published version of the manuscript.

Funding: This research received no external funding.

Acknowledgments: We thank Johanna Reck (optical analysis) and Daniel Winkler (electrical measurements), both from the Optotransmitter-Umweltschutz-Technologie e.V., for their assistance in the analysis of the SnO_2:Ta films. Markus Wollgarten from the Helmholtz-Zentrum Berlin is acknowledged for performing the transmission electron microscopy investigation.

Conflicts of Interest: All authors declare no conflict of interest.

References

1. Hosono, H.; Kikuchi, N.; Ueda, N.; Kawazoe, H. Working hypothesis to explore novel wide band gap electrically conducting amorphous oxides and examples. *J. Non Cryst. Solids* **1996**, *198*, 165–169. [CrossRef]
2. Hosono, H. Recent progress in transparent oxide semiconductors: Materials and device application. *Thin Solid Films* **2007**, *515*, 6000–6014. [CrossRef]
3. Ellmer, K. Past achievements and future challenges in the development of optically transparent electrodes. *Nat. Photonics* **2012**, *6*, 809–817. [CrossRef]
4. Vossen, J.L.; Kern, W. (Eds.) *Thin Film Processes II*; Academic Press: Cambridge, MA, USA, 1991; p. 866.
5. Lungwitz, F.; Escobar-Galindo, R.; Janke, D.; Schumann, E.; Wenisch, R.; Gemming, S.; Krause, M. Transparent conductive tantalum doped tin oxide as selectively solartransmitting coating for high temperature solar thermal applications. *Solar Ener. Mat. Solar Cells* **2019**, *196*, 84–93. [CrossRef]
6. Kavan, L.; Steier, L.; Grätzel, M. Ultrathin Buffer Layers of SnO_2 by Atomic Layer Deposition: Perfect Blocking Function and Thermal Stability. *J. Phys. Chem. C* **2017**, *121*, 342–350. [CrossRef]
7. Fischer, A. Dünne Halbleiterschichten auf Glas. *Z. Naturforsch.* **1954**, *9*, 508–511. [CrossRef]
8. Ishiguro, K.; Sasaki, T.; Arai, T.; Imai, I. Optical and Electrical Properties of Tin Oxide Films. *J. Phys. Soc. Jap.* **1958**, *13*, 296–304. [CrossRef]
9. Shanthi, E.; Banerjee, A.; Dutta, V.; Chopra, K.L. Electrical and Optical Properties of Undoped and Antimony-Doped Tin Oxide Films. *J. Appl. Phys.* **1980**, *51*, 6243–6251. [CrossRef]
10. Spence, W. The uv Absorption Edge of Tin Oxide Thin Films. *J. Appl. Phys.* **1967**, *38*, 3767–3770. [CrossRef]
11. Weissmantel, C.; Fiedler, O.; Hecht, G.; Reisse, G. Ion beam sputtering and its application for the deposition of semiconducting films. *Thin Solid Films* **1972**, *13*, 359–366. [CrossRef]
12. Manifacier, J.C. Thin Metallic Oxides as Transparent Conductors. *Thin Solid Films* **1982**, *90*, 297–308. [CrossRef]

13. Stjerna, B.; Granqvist, C.G.; Seidel, A.; Häggström, L. Characterization of RF-Sputtered SnO_x Thin Fims by Electron Microscopy, Hall-Effect Measurement, and Mössbauer Spectrometry. *J. Appl. Phys.* **1990**, *68*, 6241–6245. [CrossRef]
14. Nakao, S.; Hirose, Y.; Fukumura, T.; Hasegawa, T. Carrier generation mechanism and effect of tantalum-doping in transparent conductive amorphous SnO_2 thin films. *Jap. J. Appl. Phys.* **2014**, *53*, 5. [CrossRef]
15. Goodchild, R.G.; Webb, J.B.; Williams, D.F. Electrical Properties of Highly Conducting and Transparent Thin Films of Magnetron Sputtered SnO2. *J. Appl. Phys.* **1985**, *57*, 2308–2310. [CrossRef]
16. *Handbook of Transparent Conductors*; Ginley, D.S.; Hosono, H.; Paine, D.C. (Eds.) Springer: New York, NY, USA, 2010; p. 534.
17. Welzel, T.; Ellmer, K. Negative oxygen ion formation in reactive magnetron sputtering processes for transparent conductive oxides. *J. Vac. Sci. Techn. A* **2012**, *30*, 61306. [CrossRef]
18. Ellmer, K.; Welzel, T. Reactive Magnetron Sputtering of Transparent Conductive Oxide Thin Films: Role of Energetic Particle (Ion) Bombardment. *J. Mat. Res.* **2012**, *27*, 765–779. [CrossRef]
19. Schleife, A.; Varley, J.B.; Fuchs, F.; Rödl, C.; Bechstedt, F.; Rinke, P.; Janotti, A.; Walle, C.G.V.d. Tin dioxide from first principles: Quasiparticle electronic states and optical properties. *Phys. Rev. B* **2011**, *83*, 35116. [CrossRef]
20. Ágoston, P.; Albe, K.; Nieminen, R.M.; Puska, M.J. Intrinsic n-Type Behavior in Transparent Conducting Oxides: A Comparative Hybrid-Functional Study of In_2O_3, SnO_2, and ZnO. *Phys. Rev. Lett.* **2009**, *103*, 245501. [CrossRef]
21. Minami, T. Transparent Conducting Oxide Semiconductors for Transparent Electrodes. *Semicond. Sci. Techn.* **2005**, *20*, S35–S44. [CrossRef]
22. Ellmer, K. Resistivity of Polycrystalline Zinc Oxide Films: Current Status and Physical Limit. *J. Phys. D Appl. Phys.* **2001**, *34*, 3097–3108. [CrossRef]
23. Ellmer, K.; Bikowski, A. Topical Review: Intrinsic and Extrinsic Doping of ZnO and ZnO Alloys. *J. Phys. D* **2016**, *49*, 413002. [CrossRef]
24. Bellingham, J.R.; Phillips, W.A.; Adkins, C.J. Intrinsic Performance Limits in Transparent Conduction Oxides. *J. Mat. Sci. Lett.* **1992**, *11*, 263–265. [CrossRef]
25. Chopra, K.L.; Major, S.; Pandya, D.K. Transparent Conductors- A Status Review. *Thin Solid Films* **1983**, *102*, 1–46. [CrossRef]
26. Hartnagel, H.L.; Dawar, A.L.; Jain, A.K.; Jagadish, C. *Semiconducting Transparent Thin Films*; Institute of Physics Publishing: Bristol, UK, 1995; p. 358.
27. Swallow, J.E.N.; Williamson, B.A.D.; Whittles, T.J.; Birkett, M.; Featherston, T.J.; Peng, N.; Abbott, A.; Farnworth, M.; Cheetham, K.J.; Warren, P.; et al. Self-Compensation in Transparent Conducting F-Doped SnO_2. *Adv. Funct. Mater.* **2017**, *28*, 1701900. [CrossRef]
28. Koch, H. Zum optischen Verhalten halbleitender Zinndioxydschichten im nahen Ultrarot bei Zimmertemperatur. *phys. stat. sol.* **1963**, *3*, 1619–1628. [CrossRef]
29. Thornton, J.A. High Rate Sputtering Techniques. *Thin Solid Films* **1981**, *80*, 1–11. [CrossRef]
30. Ellmer, K. Magnetron Discharges for Thin Film Deposition. In *Low Temperature Plasmas. Fundamentals, Technologies and Techniques*; Hippler, R., Kersten, H., Schmidt, M., Schoenbach, K.H., Eds.; Wiley-VCH: Berlin/Heidelberg, Germany, 2008; Volume 2, pp. 675–715.
31. Kim, Y.-W.; Lee, S.W.; Chen, H. Microstructural evolution and electrical property of Ta-doped SnO_2 films grown on Al_2O_3(0001) by metalorganic chemical vapor deposition. *Thin Solid Films* **2002**, *405*, 256–262. [CrossRef]
32. Toyosaki, H.; Kawasaki, M.; Tokura, Y. Electrical properties of Ta-doped SnO_2 thin films epitaxially grown on TiO_2 substrate. *Appl. Phys. Lett.* **2008**, *93*, 132109. [CrossRef]
33. Nakao, S.; Yamada, N.; Hitosugi, T.; Hirose, Y.; Shimada, T.; Hasegawa, T. High Mobility Exceeding 80 $cm^2\ V^{-1}s^{-1}$ in Polycrystalline Ta-Doped SnO_2 Thin Films on Glass Using Anatase TiO_2 Seed Layers. *Appl. Phys. Express* **2010**, *3*, 031102. [CrossRef]
34. Weidner, M.; Jia, J.; Shigesato, Y.; Klein, A. Comparative study of sputter-deposited SnO_2 films doped with antimony or tantalum. *Phys. Stat. Sol* **2016**, *253*, 923–928. [CrossRef]
35. Muto, Y.; Oka, N.; Tsukamoto, N.; Iwabuchi, Y.; Kotsubo, H.; Shigesato, Y. High-rate deposition of Sb-doped SnO_2 films by reactive sputtering using the impedance control method. *Thin Soild Films* **2011**, *520*, 1178–1181. [CrossRef]

36. Jousse, D.; Constantino, C.; Chambouleyron, I. Highly Conductive and Transparent Amorphous Tin Oxide. *J. Appl. Phys.* **1983**, *54*, 431–434. [CrossRef]
37. Nomura, K.; Ohta, H.; Takagi, A.; Kamiya, T.; Hirano, M.; Hosono, H. Room-Temperature Fabrication of Transparent Flexible Thin-Film Transistors Using Amorphous Oxide Semiconductors. *Nature* **2004**, *432*, 488–492. [CrossRef] [PubMed]
38. Buchholz, D.B.; Ma, Q.; Alducin, D.; Ponce, A.; Jose-Yacaman, M.; Khanal, R.; Medvedeva, J.E.; Chang, R.P.H. The Structure and Properties of Amorphous Indium Oxide. *Chem. Mater.* **2014**, *26*, 5401–5411. [CrossRef]
39. Hu, S.; Lewis, N.S.; Ager, J.W.; Yang, J.; McKone, J.R.; Strandwitz, N.C. Thin-Film Materials for the Protection of Semiconducting Photoelectrodes in Solar-Fuel Generators. *J. Phys. Chem. C* **2015**, *115*, 24201–24228. [CrossRef]
40. Mayer, M. Ion Beam Analysis of Rough Thin Films. *Nucl. Instr. Meth. Phys. Res. B Nucl.* **2002**, *194*, 177–186. [CrossRef]
41. Bikowski, A.; Welzel, T.; Ellmer, K. The correlation between the radial distribution of high-energetic ions and the structural as well as electrical properties of magnetron sputtered ZnO:Al films. *J. Appl. Phys.* **2013**, *114*, 223716. [CrossRef]
42. Sigmund, P. Theory of Sputtering Yield of Amorphous and Polycrystalline Targets. *Phys. Rev.* **1969**, *184*, 383–416. [CrossRef]
43. Yeom, G.Y.; Kushner, M.J. Cylindrical Magnetron Discharges. I. Current-Voltage Characteristics for DC- and RF-Driven Discharge Sources. *J. Appl. Phys.* **1989**, *65*, 3816–3824. [CrossRef]
44. Ellmer, K. Magnetron Sputtering of Transparent Conductive Zinc Oxide: Relation between the Sputtering Parameters and the Electronic Properties. *J. Phys. D Appl. Phys.* **2000**, *33*, R17–R32. [CrossRef]
45. Ruscic, B.; Feller, D.; Peterson, K.A. Active Thermochemical Tables: Dissociation energies of several homonuclear first-row diatomics and related thermochemical values. *Theor. Chem. Acc.* **2014**, *133*, 1415. [CrossRef]
46. Kaufman, F. N_20 Bond Dissociation Energy. *J. Chem. Phys.* **1967**, *46*, 2449. [CrossRef]
47. Ellmer, K.; Cebulla, R.; Wendt, R. Characterization of a Magnetron Sputtering Discharge with Simultaneous RF-and DC-Excitation of the Plasma for the Deposition of Transparent and Conducting ZnO:Al-Films. *Surf. Coat. Techn.* **1998**, *98*, 1251–1256. [CrossRef]
48. Nomura, K.; Kamiya, T.; Ohta, H.; Ueda, K.; Hirano, M.; Hosono, H. Carrier Transport in Transparent Oxide Semiconductor with Intrinsic Structural Randomness Probed Using Single-Crystalline $InGaO_3(ZnO)_5$ Films. *Appl. Phys. Lett.* **2004**, *85*, 1993–1995.
49. Thienprasert, J.T.; Rujirawat, S.; Klysubun, W.; Duenow, J.N.; Coutts, T.J.; Zhang, S.B.; Look, D.C.; Limpijumnong, S. Compensation in Al-Doped ZnO by Al-Related Acceptor Complexes: Synchrotron X-Ray Absorption Spectroscopy and Theory. *Phys. Rev. Lett.* **2013**, *110*, 055502. [CrossRef]
50. Bikowski, A.; Zajac, D.A.; Vinnichenko, M.; Ellmer, K. Evidence for the AlZn-Oi defect-complex model for magnetron-sputtered aluminium-doped zinc oxide: A combined X-ray absorption near edge spectroscopy, X-ray diffraction and electronic transport study. *J. Appl. Phys.* **2019**, *126*, 045106. [CrossRef]
51. Tesmer, J.R.; Nastasi, M. (Eds.) *Handbook of Modern Ion Beam Materials Analysis*; MRS: Pittsburgh, PA, USA, 1995; p. 704.
52. Winters, H.F.; Kay, E. Gas Incorporation into Sputtered Films. *J. Appl. Phys.* **1967**, *38*, 3928–3934. [CrossRef]
53. Winters, H.F. Elementary Processes at Solid Surfaces Immersed in Low Pressure Plasmas. In *Plasma Chemistry III*; Veprek, S., Venugopalan, M., Eds.; Springer: Berlin/Heidelberg, Germany, 1980; Volume 94, pp. 69–125.
54. Lee, W.W.Y.; Oblas, D. Argon entrapment in metal films by dc triode sputtering. *J. Appl. Phys.* **1975**, *46*, 1728–1732. [CrossRef]
55. Thornton, J.A.; Hoffman, D.W. Internal Stresses in Amorphous Silicon Films Deposited by Cylindrical Magnetron Sputtering Using Ne, Ar, Kr, Xe, and Ar+H_2. *J. Vac. Sci. Techn.* **1981**, *18*, 203–207. [CrossRef]
56. Nie, M.; Bikowski, A.; Ellmer, K. Microstructure evolution of Al-doped zinc oxide and Sn-doped indium oxide deposited by radio-frequency magnetron sputtering: A comparison. *J. Appl. Phys.* **2015**, *117*, 155301. [CrossRef]
57. Minami, T.; Nanto, H.; Takata, S. Highly Conducting and Transparent SnO_2 Thin Films Prepared by RF Magnetron Sputtering on Low-Temperature Substrates. *Jap. J. Appl. Phys.* **1988**, *27*, L287–L289. [CrossRef]
58. Brousseau, J.-L.; Bourque, H.; Tessier, A.; Leblanc, R.M. Electrical properties and topography of SnO_2 thin films prepared by reactive sputtering. *Appl. Surf. Sci.* **1997**, *108*, 351–358. [CrossRef]

59. Fuchs, K. The conductivity of thin metallic films according to the electron theory of metals. *Proc. Cambridge Phil. Soc.* **1938**, *11*, 100–108. [CrossRef]
60. Sondheimer, E.H. The Mean Free Path of Electrons in Metals. *Adv. Phys.* **1952**, *1*, 1–42. [CrossRef]
61. Ellmer, K.; Mientus, R.; Weiß, V.; Rossner, H. In situ Energy-Dispersive X-Ray Diffraction System for Time-Resolved Thin Film Growth Studies. *Meas. Sci. Techn.* **2003**, *14*, 336–345. [CrossRef]
62. Kersten, H.; Deutsch, H.; Steffen, H.; Kroesen, G.M.W.; Hippler, R. The Energy Balance at Substrate Surfaces During Plasma Processing. *Vacuum* **2001**, *63*, 385–431. [CrossRef]
63. Weise, M.; Seeger, S.; Harbauer, K.; Welzel, T.; Ellmer, K. A multifunctional plasma and deposition sensor for the characterization of plasma sources for film deposition and etching. *J. Appl. Phys.* **2017**, *122*, 044503. [CrossRef]
64. Minami, T.; Oda, J.-I.; Nomoto, J.-I.; Miyata, T. Effect of Target Properties on Transparent Conducting Impurity-Doped ZnO Thin Films Deposited by DC Magnetron Sputtering. *Thin Solid Films* **2010**, *519*, 385–390. [CrossRef]
65. Kluth, O.; Schöpe, G.; Rech, B.; Menner, R.; Oertel, M.; Orgassa, K.; Schock, H.W. Comparative Material Study on RF and DC Magnetron Sputtered ZnO:Al Films. *Thin Solid Films* **2006**, *502*, 311–316. [CrossRef]
66. Szyszka, B. Magnetron Sputtering of ZnO Films. In *Transparent Conductive Zinc Oxide: Basics and Application in Thin Film Solar Cells*; Ellmer, K., Klein, A., Rech, B., Eds.; Springer: Berlin/Heidelberg, Germany, 2008; pp. 187–233.
67. Hotop, H.; Lineberger, W.C. Binding Energies in Atomic Negative Ions. *J. Phys. Chem. Ref. Data* **1975**, *4*, 539–576. [CrossRef]
68. Cuomo, J.J.; Gambino, R.J.; Harper, J.M.E.; Kuptsis, J.D. Origin and Effects of Negative Ions in the Sputtering of Intermetallic Compounds. *IBM J. Res. Dev.* **1977**, *21*, 580–583. [CrossRef]
69. Burstein, E. Anomalous Optical Absorption Limit in InSb. *Phys. Rev.* **1954**, *93*, 632–633. [CrossRef]
70. Moss, T.S. The Interpretation of the Properties of Indium Antimonide. *Proc. Phys. Soc. Sec. B* **1954**, *67*, 775–782. [CrossRef]

 MDPI

Article

Indium Tin Oxide Thin Film Deposition by Magnetron Sputtering at Room Temperature for the Manufacturing of Efficient Transparent Heaters

Jago Txintxurreta [1], Eva G-Berasategui [2,*], Rocío Ortiz [2], Oihane Hernández [2], Lucía Mendizábal [2] and Javier Barriga [2]

[1] Fagor Electrónica S. Coop. Barrio San Andrés s/n, 20500 Mondragón, Spain; jtxintxurreta@fagorelectronica.es
[2] Department of Physics of Surfaces and Materials, TEKNIKER, Basque Research and Technology Alliance (BRTA), Iñaki Goenaga 5, 20600 Eibar, Spain; rocio.ortiz@tekniker.es (R.O.); oihane.hernandez@tekniker.es (O.H.); lucia.mendizabal@tekniker.es (L.M.); javier.barriga@tekniker.es (J.B.)
* Correspondence: eva.gutierrez@tekniker.es; Tel.: +34-636993217

Abstract: Indium tin oxide (ITO) thin films are widely used as transparent electrodes in electronic devices. Many of those electronic devices are heat sensitive, thus their manufacturing process steps should not exceed 100 °C. Manufacturing competitive high-quality ITO films at low temperature at industrial scale is still a challenge. Magnetron sputtering technology is the most suitable technology fulfilling those requirements. However, ITO layer properties and the reproducibility of the process are extremely sensitive to process parameters. Here, morphological, structural, electrical, and optical characterization of the ITO layers deposited at low temperature has been successfully correlated to magnetron sputtering process parameters. It has been demonstrated that the oxygen flow controls and influences layer properties. For oxygen flow between 3–4 sccm, high quality crystalline layers were obtained with excellent optoelectronic properties (resistivity $<8 \times 10^{-4}$ Ω·cm and visible transmittance >80%). The optimized conditions were applied to successfully manufacture transparent ITO heaters on large area glass and polymeric components. When a low supply voltage (8 V) was applied to transparent heaters (THs), de-icing of the surface was produced in less than 2 min, showing uniform thermal distribution. In addition, both THs (glass and polycarbonate) showed a great stability when exposed to saline solution.

Keywords: ITO thin films; magnetron sputtering; low temperature deposition; oxygen flow; microstructure; optoelectronic properties; transparent heaters

Citation: Txintxurreta, J.; G-Berasategui, E.; Ortiz, R.; Hernández, O.; Mendizábal, L.; Barriga, J. Indium Tin Oxide Thin Film Deposition by Magnetron Sputtering at Room Temperature for the Manufacturing of Efficient Transparent Heaters. *Coatings* **2021**, *11*, 92. https://doi.org/10.3390/coatings11010092

Received: 23 December 2020
Accepted: 11 January 2021
Published: 15 January 2021

1. Introduction

Transparent conductive oxides (TCOs) have attracted wide interest due to their high optical transmittance in the visible wavelength region combined with high electrical conductivity. Due to these properties, they are extensively used as low emissivity layers in architectural glass or as transparent electrodes in multiple devices such as flat panel displays, electrochromic devices, photovoltaic cells, and organic light emitting diodes [1,2], or more recently, in microwave and radio frequency shielding devices [3]. In addition, TCO coatings can be applied as transparent heaters (THs) [4] to fast and reliably heat glass and plastic components in the automotive, locomotive, and aircraft industries (in devices such as windscreens or car headlights) to provide them with de-fogging and/or de-icing properties within harsh environments, improving the performance of currently applied solutions. For example, contemporary car windshields are laminated with polyvinyl butyral (PVB) polymeric foils containing tungsten microwires as heating element. However, these types of heated windshields show a lack in the homogeneity of heat distribution over the windshield and in their transparency. Lenses in new automotive LED headlights also lead to issues with fogging or freezing because of the condensation occurring in the interior of the outer lens. Generally, the headlight housings have vent holes with filters to recirculate air and avoid condensation, of which the number and position must be frequently modified after headlight manufacturing to optimize air recirculation, in a difficult and expensive process [5]. This has become a challenging issue from a visibility and a safety standpoint for many original equipment manufacturers (OEMs). As in the case of windshields, microwires are also commonly used as heating elements, interfering with radio detection and ranging (RADAR) and light imaging detection and ranging (LIDAR) car systems. TCO coatings can improve the performance of conventional heating elements, in terms of demonstrating a high heating power capacity with fast control of temperature and small thermal inertia without detriment to their optical transmission [6].

Indium tin oxide (ITO) is the most widely used TCO because it has unique set of properties; such as high ultraviolet absorption, high infrared reflectance, high microwave attenuation, wide bandgap (3.5–4.2 eV), high visible transmission, low electrical resistivity, good mechanical strength and abrasion resistance, chemical stability [7] and compatibility with fine patterning processes [8]. There are many deposition techniques to obtain high quality ITO films, such as radio frequency (RF) [9] and direct-current (DC) magnetron sputtering [10], E-beam evaporation [11], pulsed laser deposition (PLD) [12], and spray pyrolysis [13]. Among all of them, pulsed DC magnetron sputtering is the most suitable to manufacture ITO layers at industrial scale due to the high deposition rate and quality control of the thin films [14].

ITO layer properties are very dependent on sputtering process parameters such as temperature, pressure, target to subtract distance, discharge power and frequency, and oxygen and argon pressure during deposition, which are directly related to the physical nature of the films [15]. This physical nature encompasses structural characteristic, crystallinity, impurity levels (or doping), defect characteristics, uniformity, and stoichiometry. Thorough understanding of the relationship between sputtering process parameters, layer properties, and the physical nature of the layers is essential to obtain good reproducibility of the film properties, above all, at an industrial scale.

Many studies reporting on ITOs deposited by DC magnetron sputtering have shown that ITO thin films could reach high transparency in the visual region (90%) and high conduction properties ($\rho = 2 \times 10^{-4}$ Ω·cm). These optimal results are obtained at high substrate temperatures (>200 °C) during deposition or by post-annealing process afterwards, because temperature promotes crystallization of the layers and oxygen-vacancy creation, the main conduction mechanisms in ITO layers [16]. Many of those studies encompass the analysis of the relationship between process parameters and optoelectronic properties for high temperature ITO layers [17–20]. However, reaching high values of transmission and conductivity for ITO films at room temperature deposition is still a challenge for industrial reproducible magnetron sputtering processes. For room temperature sputtering deposition, oxygen flow effects on optical (average transmission in the visible light region, transmission and absorption spectra) and electrical (amount and mobility of electric carriers) features are two of the main factors affecting layer properties and have been previously studied, as well as effects on microstructure (grade of crystallinity, growth orientation, lattice parameter and lattice stress, grain size and structure), mainly for laboratory sputtering equipment [14,21,22]. However, there is still a lack of understanding of the relationship of those parameters for pulsed DC magnetron sputtering industrial processes at room temperature. Although TCOs were applied as THs before 1995, they have been mainly studied in the framework of industrial research and development, and very few reports can be found in the literature about the application of ITO as THs [23,24]. In these reports, ITO nanoparticles were deposited on glass substrates by spin coating, producing ITO films with high transparency but high sheet resistances (above 300 Ω/sq) even after applying annealing at high temperatures.

In the present work, a detailed study of the influence of the oxygen flow on ITO layer properties has been performed for an unbalanced pulsed DC magnetron sputtering process at low temperature. The unbalance DC magnetron sputtering process is optimum for industrial applications because it presents higher deposition rates than the balance magnetron sputtering process. However, special attention must be given to avoid the bombardment of the growing film with ion species from the plasma (O_2^- in this case) that negatively affects the microstructure of the layers and damages the electrical properties [25]. The optimization of the ITO layer at room temperature has been performed for this unbalanced magnetron sputtering process for the deposition of ITO layers at semi-industrial scale. A detailed study including microstructure analysis of ITO film series deposited under different oxygen flows has been performed. Microstructure evolution with oxygen is explained and correlated with optoelectrical properties of each film. In addition, a cost-effective process was developed and optimized for the manufacturing of THs with excellent optoelectronic properties on large area glass and polycarbonate (PC) sheets by means of conventional sputtering at room temperature.

2. Materials and Methods

ITO layers have been deposited in an industrial pulsed DC unbalanced magnetron sputtering equipment, FASTCOAT, designed and manufactured by TEKNIKER (Eibar, Spain). This equipment has one unbalanced magnetron of 550 × 125 mm^2 target size. The ITO (In_2O_3:Sn_2O_3 at. % 90:10 99.99% purity) target was placed 120 mm from a rotatable substrate holder. Two types of substrates were used for characterization purpose, silicon wafers and microscope glass slides (Menzel-Gläser). The ITO films were deposited at 1500 W average power, using an Advanced Energy DC Pinnacle Plus (Advance Energy, Denver, CO, USA) power supply, under the following pulsing parameters: 75 kHz pulse frequency, 4 µs pulse-off time, and a duty cycle of 70%. The vacuum chamber, which had a 230 L volume, was pumped down with a pumping speed of 1200 L/s to a base pressure of 2 × 10^{-6} mbar before the deposition. The gas entry supply was located closer to the substrate than to the target to avoid the acceleration of the oxygen ions by the potential applied to the target and, therefore, the bombardment of the growing film with O^+ species from the plasma. During the deposition, argon flow was maintained constant at 150 sccm, while variable oxygen flow was introduced into the chamber by mass flow controllers, resulting in 1.8 × 10^{-3} mbar process pressure. Before deposition, the target was pre-sputtered for 10 min; the first 5 min with the same Ar flow of the process but without oxygen, and the last 5 min with both O_2 and Ar flows applied during the process.

Oxygen flow was varied from 0 to 6 sccm, testing 7 different values, to correlate the oxygen flow influence with layer properties and morphology. The deposition rate of the layers was 20 nm/min (remaining unchanged for the different oxygen flow applied), obtaining 140 nm thick layers by depositing ITO for 7 min. Vacuum time and base pressure were kept constant for all the deposition processes performed at different oxygen flow, so the possible presence of residual impurities in the films caused by the gettering of water vapor from the chamber was the same.

The electrical properties were measured by 4-point probe [26] and Hall Effect measurements (ECOPIA, Anyang City, South Korea). Optical transmittance was measured with a Perkin Elmer Lambda UV/VIS/NIR spectrophotometer (Perkin Elmer, Waltham, MA, USA). Morphology studies were made by an ULTRA Plus Carl-Zeiss field emission scanning electron microscope (FE SEM, Carl-Zeiss, Oberkochen, Germany) and an atomic force microscope Solver PRO NT-MDT (NT-MDT SI, Limerick, Ireland). The average surface roughness (S_a) and root mean square roughness (S_q) values have been calculated from 3 measurements of 3 × 3 µm^2 AFM images of ITO layers deposited with different oxygen flows. A D8 Advance Bruker X-ray diffractometer (Bruker, Billerica, MA, USA) was used to determine the crystallinity of the films deposited on glass by measuring with Cu-Kα radiation in $\theta-2\theta$ geometry with a step of 0.02°, a step time of 7.2 s, and applying a grazing incidence of 2° (GIXRD).

Glass and polycarbonate (PC) samples of 100 × 100 mm^2 were used as prototypes of THs to evaluate their performance. Contacts were made by a conductor tape (tin-plated copper foil, PPI Adhesive Products Ltd.,Waterford, Ireland) to apply a DC voltage to the prototypes, and a thermographic camera was used to measure the temperature versus time to calculate the saturation temperatures at different applied voltages (2–12 V for standard uses). Temperature stability was also analyzed by applying a constant voltage (8 V) for 4 h. In addition, prototype de-fogging and de-icing properties were examined by measuring the time required to defog and/or defrost the samples after keeping them for 30 min in a refrigerator at −21 °C and applying a DC voltage of 8 V. Prototype durability was analyzed by applying adhesion tests using scotch tape (standard MIL-C-675C [27] for the coating of glass optical elements).

3. Results

3.1. Effects of Oxygen Flow on the Morphology of ITO Thin Films

The effect of the oxygen flow on the morphology of the ITO layers is shown in Figure 1 (SEM images) and Figure 2 (AFM images). For low oxygen flows (0–1 sccm), the ITO surface was smooth (with surface roughness values around 0.8 nm) with no spikes and with a cauliflower-like microstructure. When oxygen flow increased (2–6 sccm), the surface became more granular, with greater surface roughness, and shaped crystallites appeared, increasing the number of polycrystalline grains with the oxygen flow. For an oxygen flow of 2–3 sccm, the cauliflower-like structure was mixed with crystalline grains. From 4 sccm and above, homogeneous polycrystalline surfaces were observed, with a significant increase in surface roughness (until 2 nm of S_a and 2.5–3 of S_q).

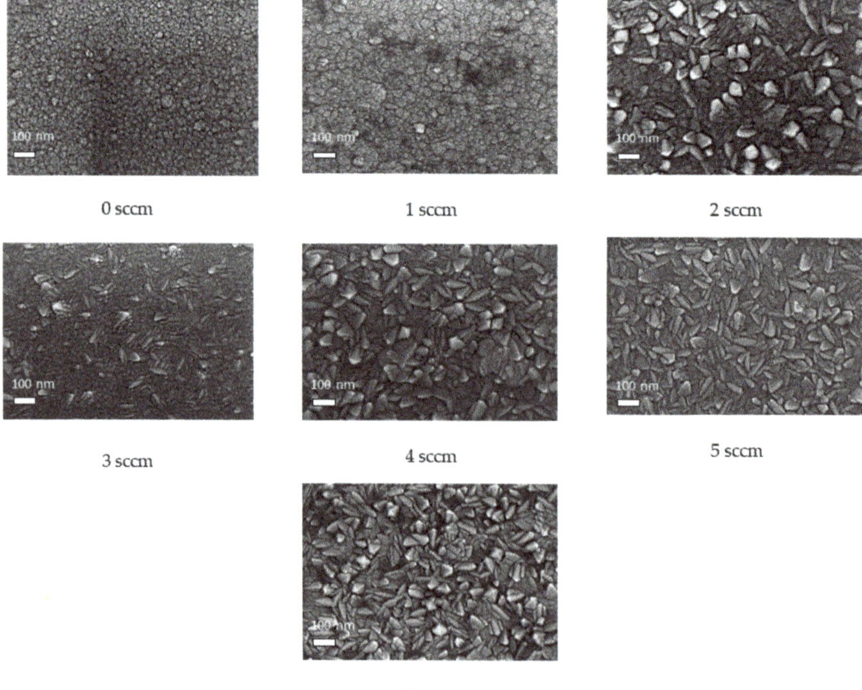

Figure 1. SEM images of the surface of indium tin oxide (ITO) layers deposited with different oxygen flows.

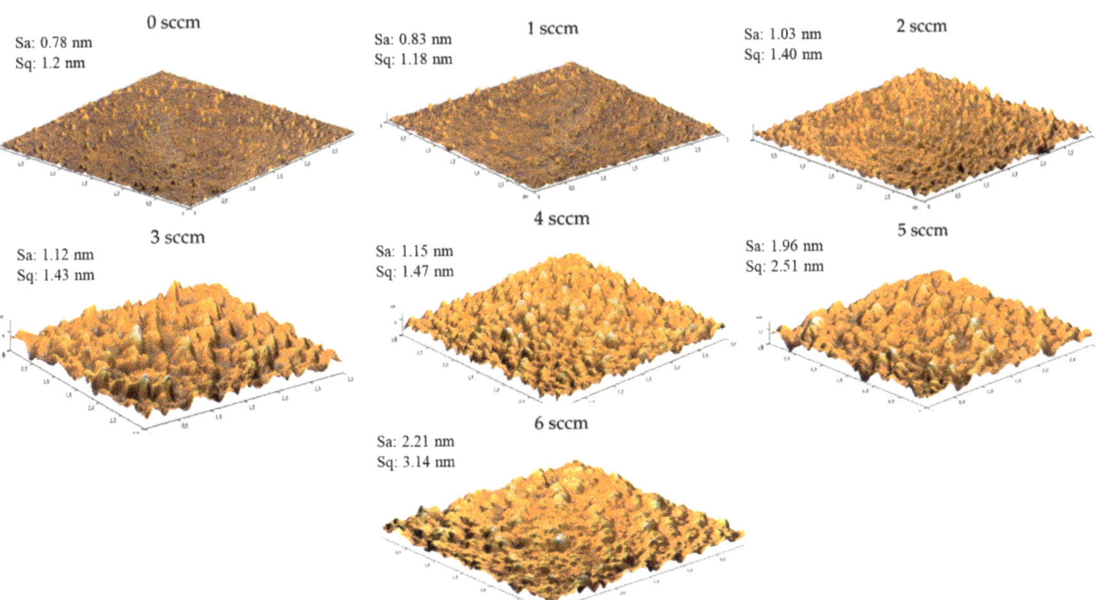

Figure 2. 2D and 3D AFM images of ITO layers surface. Average (S_a) and root mean square (S_q) roughness were calculated.

To correlate surface morphology with layer crystalline structure, XRD analysis was performed (Figure 3). XRD patterns showed that the oxygen flow affects the crystallinity of the layers. XRD pattern for the ITO deposited with 0 sccm of oxygen showed a halo pattern around 2θ = 32°–35°, which is characteristic of amorphous materials. For 1 sccm oxygen flow, a broad diffraction peak appeared around 2θ = 30°, corresponding to an incipient crystallization of the ITO layer in the (222)-oriented bcc structure of In_2O_3 [28]. Both results can be correlated with the SEM and AFM pictures of 0 and 1 sccm, where surface was smooth, and no grains were observed. When oxygen flow increased (2–5 sccm), the (222) diffraction peak became more intense and narrower, indicating the growth of the (222)-oriented crystallites, while new diffraction peaks appeared showing the formation of crystallites with other orientations. This correlates with the SEM (Figure 1) and AFM (Figure 2) surface images, where geometrically regular forms (scales and pyramidal peaks) fill the surface. However, a reduction in (222) peak intensity was observed when the oxygen flow was further increased up to 6 sccm A shift to higher angles was observed for the (222) peak when oxygen flow varied from 4 to 5 sccm. When increasing the oxygen flow from 1 sccm onwards, small diffraction peaks typical of polycrystalline ITO thin films with a cubic indium oxide structure appeared [17], showing a minor formation of crystallites with other orientations such as (400), (332), (431), (440), and (622).

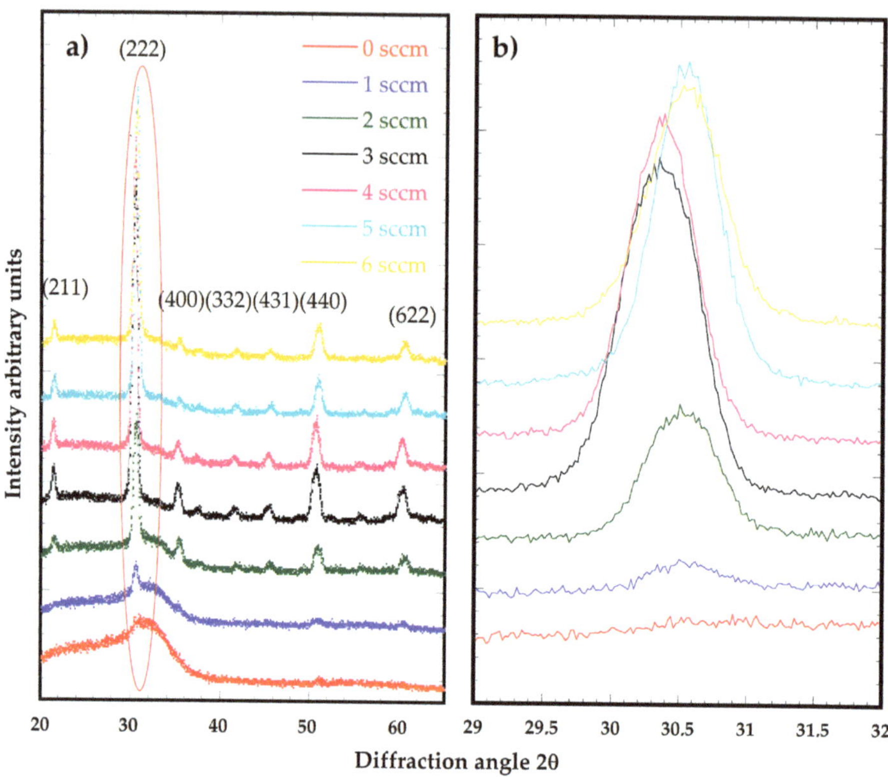

Figure 3. XRD patterns of ITO deposited with an oxygen flow from 0 to 6 sccm: (**a**) measured in diffraction angle from 20 to 65; (**b**) Zoomed view of (222) peak.

Figure 4 shows the optical transmittance of each sample from the ultraviolet (λ = 250 nm) to the near-infrared (λ = 2500 nm) region. The mean transmission values in the visible region increased with the oxygen flow. Non-crystallized ITO layers (produced with 0 and 1 sccm of oxygen flows) had much lower average light transmission (around 65%) than crystallized layers produced with an O_2 flow >3 sccm, which showed transmission values above 80% in the visible region, reaching the maximum value of 82% for 6 sccm.

In the near-infrared region (800–2500 nm), the behavior was different: first, the transmission decreased when the oxygen flow was increased from 0 to 3 sccm, whereas the opposite tendency was observed upon 4 sccm.

The evolution of electrical properties of the ITO layers, such as resistivity, carrier concentration, and Hall mobility with the oxygen flow is shown in Figure 5. We observed that the electrical properties were highly influenced by the oxygen flow applied during deposition of the ITO layer. Increasing the oxygen flow until 5 sccm boosted the Hall mobility to reach a maximum value of (36 ± 1) $cm^2/V{\cdot}s$. On the contrary, the increase in the oxygen flows led (from 0 to 2 sccm) to a slight enhancement of the carrier concentration with a subsequent reduction until reaching a minimum value below 1×10^{20} cm^{-3} at 6 sccm of oxygen flow. The resistivity seemed to be dominated by the Hall mobility, remaining at a minimum value of approximately 1×10^{-3} $\Omega{\cdot}cm$ for hall mobility above 10 $cm^2/V{\cdot}s$. In this range, only a significant increase in resistivity was produced when the carrier concentration decreased below 1×10^{20} cm^{-3} at 6 sccm of oxygen flow. It is also worth highlighting that the sharpest decrease in resistivity was produced when the ITO layer was deposited in the presence of oxygen, with respect to those obtained in pure argon atmosphere.

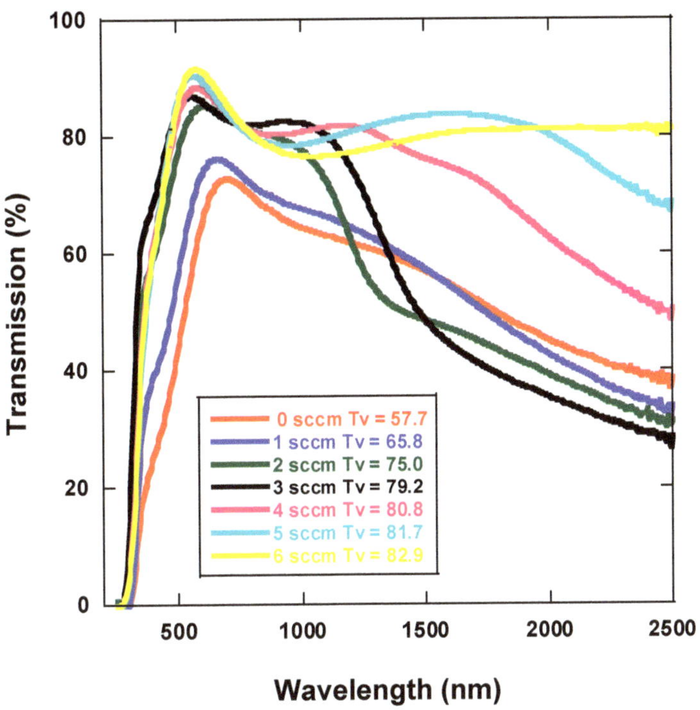

Figure 4. Transmittance of ITO films sputtered with different oxygen flows. Mean transmittance values (T_v) for each oxygen flow are given.

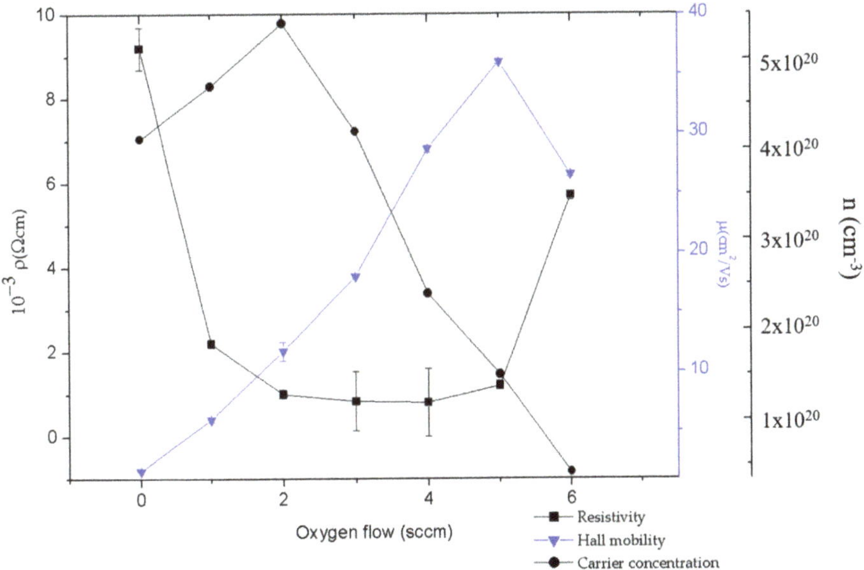

Figure 5. Variation of carrier concentration (circular dots), carrier mobility (triangular dots), and resistivity (squared dots) of ITO layers deposited with and without different oxygen flows. If error bars are not visible, they are smaller than the symbol size.

3.2. Manufacturing of Transparent ITO Heaters

Table 1 summarizes the optoelectronic properties of the deposited ITO layers. In addition to the previously measured properties, the sheet resistance (measure of resistance on uniform thin films) and the Haacke's figure of merit (FoM, a dimensionless parameter to evaluate the performance of thermoelectric materials) were also calculated because of their importance in many industrial applications of TCOs. FoM was calculated from the average optical transmittance at VIS region, and the sheet resistance (*SR*) [29]:

$$\phi_{VIS} = \frac{T_{av}}{SR} \tag{1}$$

Table 1. Average optical transmission in the visible region, *T* (%), Hall mobility (μ), carrier concentration (n), sheet resistance (*SR*), electrical resistivity (ρ), and FoM (ϕ_{VIS}) of ITO layers.

Oxygen Flow (sccm)	*T* (%)	μ (cm^2/Vs)	n (/cm^3)	*SR* (Ω/sq)	ρ (Ω·cm)	FoM ϕ_{VIS} (Ω^{-1})
0	57.7 ± 0.7	1.7 ± 0.2	$(4.12 \pm 0.03) \times 10^{20}$	680 ± 50	$(9.2 \pm 0.5) \times 10^{-3}$	0.85×10^{-3}
1	65.8 ± 0.4	5.97 ± 0.06	$(4.7 \pm 0.2) \times 10^{20}$	170 ± 10	$(2.2 \pm 0.1) \times 10^{-3}$	3.87×10^{-3}
2	75 ± 1	11.7 ± 0.4	$(5.4 \pm 0.8) \times 10^{20}$	79 ± 9	$(1.0 \pm 0.1) \times 10^{-3}$	9.49×10^{-3}
3	79.2 ± 0.9	18 ± 1	$(4.2 \pm 0.2) \times 10^{20}$	63 ± 5	$(8.3 \pm 0.7) \times 10^{-4}$	$12.6 \cdot \times 10^{-3}$
4	80.8 ± 0.3	28.7 ± 0.9	$(2.4 \pm 0.3) \times 10^{20}$	77 ± 6	$(7.9 \pm 0.8) \times 10^{-4}$	$10.5 \cdot \times 10^{-3}$
5	81.7 ± 0.6	36 ± 1	$(1.5 \pm 0.2) \times 10^{20}$	95 ± 9	$(1.2 \pm 0.1) \times 10^{-3}$	8.60×10^{-3}
6	82.9 ± 0.5	26 ± 1	$(4.2 \pm 0.1) \times 10^{19}$	428 ± 15	$(5.6 \pm 0.1) \times 10^{-3}$	$1.94 \cdot \times 10^{-3}$

The ITO thin film showing optimal optoelectronic properties, in terms of reaching a commitment between transmission and conductivity, was the one deposited with an oxygen flow of 3 sccm, which showed the highest FoM ($12.6 \times 10^{-3}\ \Omega^{-1}$). These conditions were applied on large-area glass and PC sheets of 100 mm^2 for the manufacturing of transparent ITO heaters by a cost-effective process developed at room temperature. Electrical circuits were mounted on coated glass and PC samples to study heat transmission and evaluate the coating performance.

Figure 6 shows the saturation temperature of the selected ITO thin film deposited on glass (a) and PC (b) as a function of the applied power. The thermal resistance was calculated from the slope of the curve fit.

The times required to completely defrost the ITO coated glass and PC samples when applying a DC voltage of 8 V after keeping the samples in a refrigerator at −21 °C for 30 min were 2 and 1.5 min, respectively (Figure 7). The long-term working stability of the ITO thin film was proven by applying a constant voltage of 8 V for 4 h (Figure 8). Both samples did not show any sign of degradation after applying the adhesion tape test and after being immersed for 24 h in saline solution.

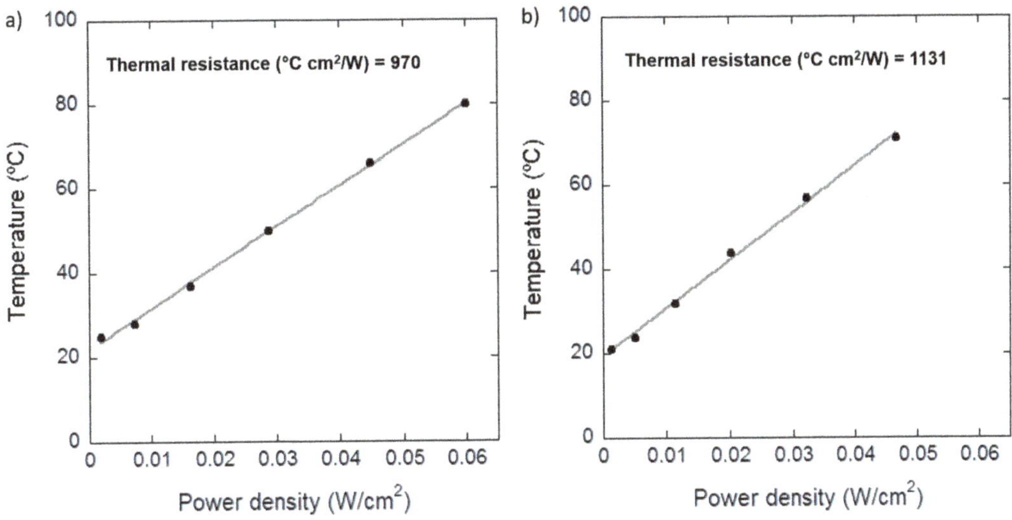

Figure 6. Saturation temperature of ITO thin films deposited on glass (**a**) and PC (**b**) samples as a function of the applied power.

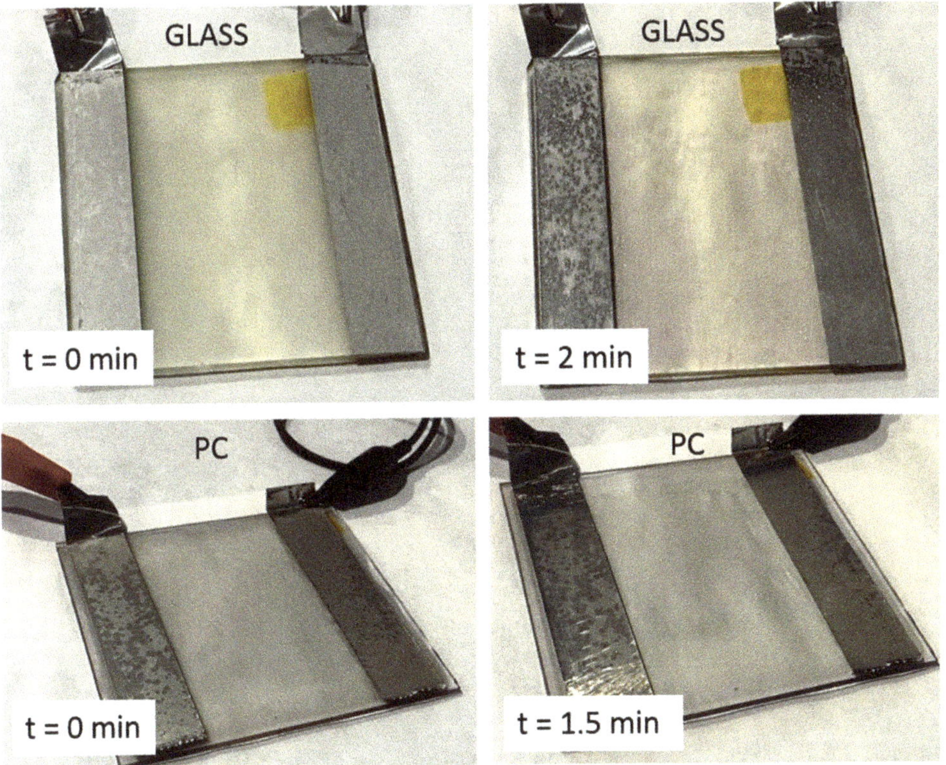

Figure 7. Glass and PC 10 × 10 cm^2 samples coated by an ITO thin film and kept in a refrigerator for 30 min at −21 °C, before and after applying a DC voltage of 8 V for 2 min.

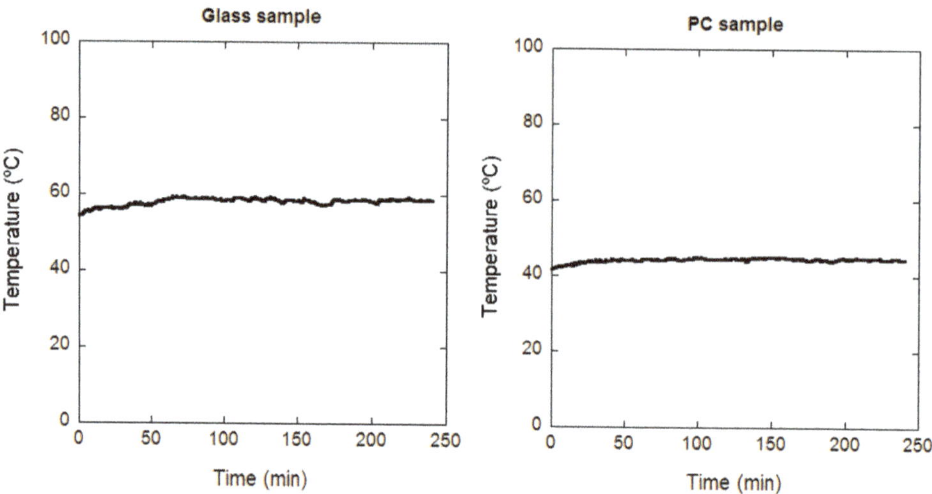

Figure 8. Temperature measured by a thermographic camera on glass and PC samples coated by ITO thin films when applying a DC voltage of 8 V for 4 h.

4. Discussion

4.1. Effects of Oxygen Flow on Microstructural and Optoelectronic Properties of ITO Thin Films

Although all samples showed very low average surface roughness (<2.5 nm), surface morphology changed drastically with increasing the oxygen flow, as previously reported by other authors [30,31]. We can divide the oxygen flow range in three different regions. In each of those regions, the increase in the amount of oxygen during deposition showed different effects in the layer morphology and microstructure, and, hence, in the optoelectronic properties of the layers. Those regions are low oxygen or suboxide regions (from 0 to 1 sccm), medium or optimum oxygen regions (from 2 to 5 sccm), and high or excessive oxygen regions (above 6 sccm).

4.1.1. Low Oxygen Region or Suboxide Region (0–1 sccm)

ITO layers in this region were mainly amorphous with very smooth surface and an average surface roughness below 1 nm. A critical oxygen flow above 1 sccm was needed to crystallize the ITO layer. Many studies have stated that there is a critical oxygen flow for the crystallization of ITO films deposited below crystallization temperature (160–180 °C) [32–34]. In these reports, the enhanced crystallization when introducing O_2 flux was attributed, on one hand, to its effect on the promotion of the formation of stoichiometric InO_3 and, on the other hand, to the production of energetic O atoms and O^- ions generated at the target surface and reaching the substrate and providing enough energy to the adatoms to induce the formation of crystalline structures. The lack of oxygen input and the consequential lack of crystallization significantly affected the optoelectronic properties of the ITO layer, showing a mean optical transmission in the visible region around 65%, usually attributed to the formation of substoichiometric compounds such as InO. High resistivity values were observed ($(9.2 \pm 0.5) \times 10^{-3}$ Ω·cm), related to the low carrier mobility (<6 $cm^2/V{\cdot}s$), which can be attributed to the higher number of carrier collisions that occur in a disordered amorphous structure.

4.1.2. Medium Oxygen—Optimum Region (2–5 sccm)

ITO crystalline layers with a (222)-oriented bcc structure were formed in this region. The increase in crystallinity caused by the higher oxygen input to the sputtering process produced a decrease in the amount of electron scatter centers (such as grain boundaries, impurities, or defects in the crystalline structure), enhancing the hall mobility (Table 1) until a maximum value of 36 $cm^2/V{\cdot}s$ for 5 sccm of oxygen flow.

It is well known that charge carriers of the ITO thin films are either contributed by Sn^{+4} ions or oxygen vacancies. The observed opposite behavior of carrier concentration when increasing the oxygen flow during deposition was due to a decrease in the number of oxygen vacancies, which are electron donors [35]. Besides, oxygen combines and neutralizes Sn^{4+}, forming Sn–O complexes and further reducing the carrier amount. The significant increase in carrier mobility while keeping the carrier concentration diminishing but in the same order of magnitude improves the conductivity of the ITO thin film, as observed by other authors in previous studies of transparent conductive electrodes [36]. A decrease in the number of carriers was also noticed in the transmission spectra in the NIR (near-infrared) region (Figure 4). Free carriers can be excited with photons which have wavelengths in this region; thus, if carrier concentration is high, absorption in that range will occur [1,37]. Therefore, ITO layers deposited with oxygen inputs which caused the highest carrier concentrations (0–3 sccm) (Table 1) showed the lowest transmission at NIR. The optical gap energy of amorphous ITO thin films increases with crystallization, increasing its optical transparency [38], which is consistent with our results.

The higher VIS transmission observed in ITO layers deposited in this region was caused by the higher crystallinity of the thin films, which produced less light scattering. The change in average transmission between 2 and 5 sccm of oxygen was lower than the change between 1 and 2 sccm (Table 1). This suggests that once the ITO layer was crystallized, the increase in the VIS transmission was more gradual for higher degrees of crystallization than when the material changed from amorphous to crystalline. A shift to higher angles was observed for the (222) peak when oxygen flow varied from 4 to 5 sccm. This shift can be attributed to the stress induced in the layer by the high oxygen flow, as observed by other authors [22,34], who have reported an increase in the measured residual compressive stress of ITO films for the highest applied oxygen flow. They stated that this compressive stress was caused by the higher plasma bombardment energy involved in film deposition with high oxygen flow, producing more dense films.

4.1.3. High Oxygen or Oxygen Excess Region (up to 6 sccm)

As mentioned before, the contribution to carrier concentration in ITO layers arises from oxygen vacancies. Low-temperature ITO deposition with high oxygen flux significantly reduced the number of oxygen vacancies in the layer, as suggested by the abrupt reduction observed in carrier concentration. According to Lee et al. [39], the excess oxygen can act like two types of scattering centers: on one hand, forming Sn^{+}–O complexes with near Sn ions to create neutral electron scattering centers and limiting the diffusion of these ions from interstitial locations and grain boundaries into the indium cation sites [37]; and, on the other hand, acting like traps to capture the electron carriers. Moreover, the carrier mobility changed its behavior and decreased when oxygen flow increased from 5 to 6 sccm, because of the higher number of scattering centers. The decrease in carrier concentration had a clear influence on the optical properties of the ITO layer. A significant increase at the NIR region was observed because of the decrease in the number of NIR-absorbing free carriers.

4.2. Manufacturing of Transparent ITO Heaters

The developed 10×10 cm^2 TH prototypes showed high optical transmission and an appropriate thermal response time (lower than 2 min). Different steady state temperatures (from 25 to 80 °C) can be reached applying low voltage (below 12 V) to adapt their performance to the requirements of the application (below 30 °C for de-icing or defogging uses, or higher for fast defrosting in automotive parts [4]). The voltage may need to be increased to heat larger parts (which involves larger distances between the electrodes) and maintains the good heating properties reached in our prototypes (200–600 W/m^2). It is worth highlighting that these THs also exhibited uniform thermal distribution over the heating area (Figure 9), which is essential for eye comfort and for avoiding the formation of hot spots that can damage the TH [40]. Both THs (glass and PC) showed a great stability when exposed to saline solution.

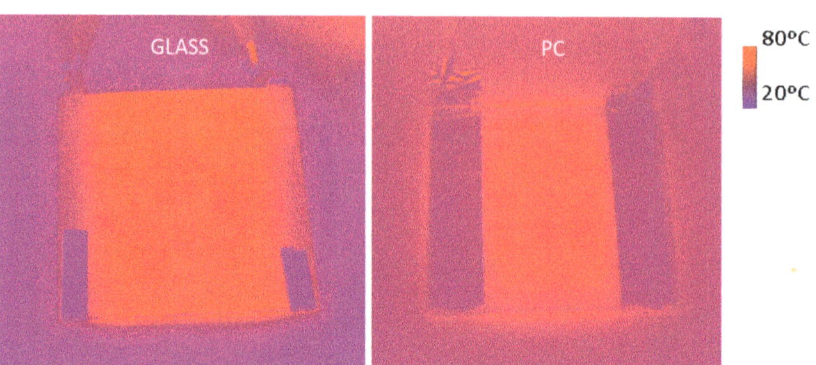

Figure 9. Thermographic image of ITO coated glass when applying a voltage of 8 V. The square-shaped area marked in the PC image indicates a region of interest selected in this case.

5. Conclusions

Magnetron sputtering is the most useful and effective technique to deposit ITO thin films. Many methods exist to obtain high quality ITO layers, but an in-depth understanding of the deposition process is essential to grow reproducible ITO films. Small changes in certain process parameters drastically alter ITO properties. In the case of low temperature-deposited ITO, oxygen input in the process is necessary to obtain high conductivity and transparent layers. As we have confirmed, the microstructure of the material is strongly dependent on the oxygen flow, changing from amorphous to crystalline ITO layers with very small variation of oxygen amount on the gas mixture. ITO films deposited above the critical oxygen flow will be crystalline, and they will show a high transparency (>80%) in the visible region. Obtaining good electrical properties is more challenging. On one hand, it depends on the microstructure features of the layers which is directly related to the mobility of the electric carriers; On the other hand, carrier concentration is proportional to the number of oxygen vacancies and Sn^{+4} ions in the microstructure. Therefore, ITO films with appropriate opto-electronic properties must be grown under specific oxygen flow conditions.

Applying the found optimal oxygen flow conditions, efficient transparent ITO heaters can be manufactured by a cost-effective and robust process on glass and polymeric components, such as those used in the automotive industry (windshields or car headlights).

Author Contributions: Conceptualization, E.G.-B., L.M., and J.B.; Methodology, E.G.-B., J.T., R.O., and O.H.; Software, J.T.; Validation, J.T., E.G.-B., and L.M; Formal Analysis, J.T., E.G.-B., and R.O.; Investigation, J.T., R.O., and O.H; Resources, J.B.; Data Curation, J.T., E.G.-B., and R.O.; Writing—Original Draft Preparation, J.T., E.G.-B., and L.M.; Writing—Review & Editing, E.G.-B., R.O., and O.H.; Visualization, J.T., E.G.-B., and R.O.; Supervision, E.G.-B., L.M., and J.B.; Project Administration, E.G.-B.; Funding Acquisition, E.G.-B., L.M., and J.B. All authors have read and agreed to the published version of the manuscript.

Funding: This research received no external funding.

Data Availability Statement: Data is contained within the article.

Conflicts of Interest: The authors declare no conflict of interest.

References

1. Granqvist, C.G.; Hultåker, A. Transparent and conducting ITO films: New developments and applications. *Thin Solid Films* **2002**, *411*, 1–5. [CrossRef]
2. Hotovy, J.; Hüpkes, J.; Böttler, W.; Marins, E.; Spiess, L.; Kups, T.; Smirnov, V.; Hotovu, I.; Kovác, J. Sputtered ITO for applications in thin-film silicon solar cells: Relationship between structural and electrical properties. *Appl. Surf. Sci.* **2013**, *269*, 81–87. [CrossRef]
3. Maniyara, R.A.; Mkhitaryan, V.K.; Chen, T.L.; Ghosh, D.S.; Pruneri, V. An antireflection transparent conductor with ultralow optical loss (<2%) and electrical resistance. *Nat. Commun.* **2016**, *7*, 13771. [CrossRef] [PubMed]
4. Papanastasiou, D.T.; Schultheiss, A.; Muñoz-Rojas, D.; Cell, C.; Carella, A.; Simonato, J.P.; Bellet, D. Transparent heaters: A review. *Adv. Funct. Mater.* **2020**, *30*, 1910225. [CrossRef]
5. Oxyzoglou, I.; Tejero, A. Prediction of condensation forming in automotive headlights using CFD. *Tech. Rep. TOYOTA Motor Eur.* **2018**. [CrossRef]
6. Szyszka, B.; Dewald, W.; Gurram, S.K.; Pflug, A.; Schulz, C.; Siemers, M.; Sittinger, V.; Ulrich, S. Recent developments in the field of transparent conductive oxide films for spectral selective coatings, electronics and photovoltaics. *Curr. Appl. Phys.* **2012**, *12*, S2–S11. [CrossRef]
7. Elhalawaty, S.; Sivaramakrishnan, K.; Theodore, N.D.; Alford, T.L. The effect of sputtering pressure on electrical optical and structure properties of indium tin oxide on glass. *Thin Solid Films* **2010**, *518*, 3326–3331. [CrossRef]
8. Park, C.H.; Lee, J.H.; Choi, B.H. Effects of the surface treatment of ITO anode layer patterned with shadow mask technology on characteristics of organic light-emitting diodes. *Org. Electron.* **2013**, *14*, 3172–3179. [CrossRef]
9. Yang, C.; Lee, S.; Lin, T.; Chen, S. Electrical and optical properties of indium tin oxide films prepared on plastic substrate by radio frequency magnetron sputtering. *Thin Solid Films* **2008**, *516*, 1984–1991. [CrossRef]
10. Sato, Y.; Taketomo, M.; Ito, N.; Miyamura, A.; Shigestao, Y. Comparative study on early stages of film growth for transparent conductive oxide films deposited by dc magnetron sputtering. *Thin Solid Films* **2008**, *516*, 4598–4602. [CrossRef]
11. Fallah, H.R.; Ghasemi, M.; Vahid, M.J. Substrate temperature effect on transparent heat reflecting nanocrystalline IOT films prepared by electron beam evaporation. *Renew. Energy* **2009**, *35*, 1527–1530. [CrossRef]
12. Viespe, C.; Nicolae, I.; Sima, C.; Grigoriu, C.; Medianu, R. ITO films deposited by advanced pulsed laser deposition. *Thin Solid Films* **2007**, *515*, 8771–8775. [CrossRef]
13. Rozati, M.; Ganj, T. Transparent conductive Sn-doped indium oxide thin film deposited by spray pyrolysis techniques. *Renew. Energy* **2004**, *29*, 1671–1676. [CrossRef]
14. Kurdesau, F.; Khripunov, G.; da Cunha, A.F.; Kaelin, M.; Tiwari, A.N. Comparative study of ITO layers deposited by DC and RF magnetron sputtering at room temperature. *J. Non-Crystall. Solids* **2006**, *352*, 1466–1470. [CrossRef]
15. Bhagwat, S.; Howson, R.P. Use of the magnetron-sputtering technique for the control of the properties of indium tin oxide thin films. *Surf. Coat. Technol.* **1999**, *111*, 163–171. [CrossRef]
16. Guillén, C.; Herrero, J. Influence of oxygen in the deposition and annealing atmosphere on the characteristic on ITO thin film prepared but sputtering at room temperature. *Thin Solid Films* **2005**, *480–481*, 129–132.
17. Guillén, C.; Herrero, J. Polycrystalline growth and recrystallinization process in sputtering ITO thin films. *Thin Solid Films* **2006**, *510*, 260–264. [CrossRef]
18. Morikawa, H.; Fujita, M. Crystallization and electrical property change on the annealing of amorphous indium-oxide and indium tin oxide films. *Thin Solid Films* **2000**, *359*, 61–67. [CrossRef]
19. Gui, Y.; Miscuglio, M.; Ma, Z.; Tahersima, M.H.; Sun, S.; Amin, R.; Dalir, H.; Sorger, V.J. Towards integrated mecatronics: A holistic approach on precise optical and electrical properties of Indium Tin Oxide. *Sci. Rep.* **2019**, *9*, 11279. [CrossRef]
20. Mudryi, A.; Ivaniukovich, A.V.; Ulyashin, A. Deposition by magnetron sputtering and characterization of indium tin oxide thin films. *Thin Solid Films* **2007**, *515*, 6489–6492. [CrossRef]
21. Ghorannevis, Z.; Akbarnejad, E.; Ghoranneviss, M. Structural and morphological properties of ITO thin films grown by magnetron sputtering. *Theor. Appl. Phys.* **2015**, *9*, 285–290. [CrossRef]
22. Tien, C.-L.; Lin, H.-Y.; Chang, C.-K.; Tang, C.-J. Effect of Oxygen Flow Rate on the Optical, Electrical, and Mechanical Properties of DC Sputtering ITO Thin Films. *Adv. Condens. Matter Phys.* **2018**, *2018*, 1–6. [CrossRef]
23. Im, K.; Cho, K.; Kim, J.; Kim, S. Transparent heaters based on solution-processed indium tin oxide nanoparticles. *Thin Solid Films* **2010**, *518*, 3960–3963. [CrossRef]
24. Kim, C.; Park, J.W.; Kim, J.; Hong, S.J.; Lee, M.J. A highly efficient indium tin oxide nanoparticles (ITO-NPs) transparent heater based on solution-process optimized with oxygen vacancy control. *J. Alloys Compd.* **2017**, *726*, 712–719. [CrossRef]
25. Moon, C.S.; Han, J.G. Low temperature synthesis of ITO thin film on polymer in Ar_2/H_2 plasm by pulsed DC magnetron sputtering. *Thin Solid Films* **2008**, *516*, 6560–6564. [CrossRef]
26. Bouroushian, M. Characterization of Thin Films by Low Incidence X-Ray Diffraction. *Cryst. Struct. Theory Appl.* **2012**, *1*, 35–39. [CrossRef]
27. Coating of Glass Optical Elements (Anti-Reflection). Military Specification, MIL-C-675C. 1980. Available online: https://www.irdglass.com/wp-content/uploads/2016/07/MIL-C-675C-AR-coatings.pdf (accessed on 14 January 2021).
28. Dutta, J.; Ray, S. Variation in structural and electrical properties of magnetron-sputtered Indium Tin Oxide films with deposition parameters. *Thin Solid Films* **1988**, *162*, 119–127. [CrossRef]
29. Haacke, G.J. New figure of merit for transparent conductors. *J. Appl. Phys.* **1976**, *47*, 4086. [CrossRef]

30. Xian, S.; Nie, L.; Qin, J.; Kang, T.; Li, C.; Xie, J.; Deng, L.; Lei, B. Effect of oxygen stoichiometry on the structure, optical and epsilon-near-zero properties of indium tin oxide films. *Opt. Express* **2019**, *27*, 28618–28628. [CrossRef]
31. Munir, M.M.; Iskandar, F.; Yun, K.M.; Okuyama, K.; Abdullah, M. Optical and electrical properties of indium tin oxide nanofibers prepared by electrospinning. *Nanotechnology* **2008**, *19*, 145603. [CrossRef]
32. Buchanan, M.; Webb, J.B.; Williams, D.F. Preparation of conducting and transparent thin films of tin-doped indium oxide by magnetron sputtering. *Appl. Phys. Lett.* **1980**, *37*, 213–215. [CrossRef]
33. Hoshi, Y.; Kato, H.; Funatsu, K. Structure and electrical properties of ITO thin films deposited at high rate by facing target sputtering. *Thin Solid Films* **2003**, *445*, 245–250. [CrossRef]
34. Vink, T.J.; Walrave, W.; Daams, J.L.C.; Baarslag, P.C.; van den Meerakker, J.E.A.M. On the homogeneity of sputter-deposited ITO films Part I. Stress and microstructure. *Thin Solid Films* **1995**, *266*, 145–151. [CrossRef]
35. Fan, J.C.C.; Goodenough, J.B. X-ray photoemission spectroscopy studies of Sn-doped indium-oxide films. *J. Appl. Phys.* **1977**, *48*, 3524–3531. [CrossRef]
36. Masis, M.M.; De Wolf, S.; Woods-Robinson, R.; Ager, J.W.; Ballif, C. Transparent Electrodes for Efficient Optoelectronic. *Adv. Electron. Mater.* **2017**, *3*, 1600529. [CrossRef]
37. Wong, F.L.; Fung, M.K.; Tong, S.W.; Lee, C.S.; Lee, S.T. Flexible organic light-emitting device based on magnetron sputtered indium-tin-oxide on plastic substrate. *Thin Solid Films* **2004**, *466*, 225–230. [CrossRef]
38. Kim, J.; Shrestha, S.; Souri, M.; Connell, J.G.; Park, S.; Seo, A. High-temperature optical properties of indium tin oxide thin-films. *Nat. Res. Sci. Rep.* **2020**, *10*, 12486. [CrossRef]
39. Lee, H.C.; Ok Park, O. Behaviours of carrier concentrations and mobilities in indium-tin oxide thin films by DC magnetron sputtering at various oxygen flow rates. *Vacuum* **2004**, *77*, 69–77. [CrossRef]
40. Khaligh, H.H.; Xu, L.; Khosropour, A.; Madeira, A.; Romano, M.; Pradére, C.; Tréguer-Delapierre, M.; Servant, L.; Pope, M.A.; Goldthorpe, I.A. The Joule heating problem in silver nanowire transparent electrodes. *Nanotechnology* **2017**, *28*, 425703. [CrossRef]

Article

Tailoring Crystalline Structure of Titanium Oxide Films for Optical Applications Using Non-Biased Filtered Cathodic Vacuum Arc Deposition at Room Temperature

Elena Guillén [1], Matthias Krause [2], Irene Heras [3], Gonzalo Rincón-Llorente [4] and Ramón Escobar-Galindo [5,*]

1 Profactor GmbH, Im Stadtgut A2, 4407 Steyr-Gleink, Austria; elena.guillen@profactor.at
2 Helmholtz-Zentrum Dresden-Rossendorf, Bautzner Landstraße 400, 01328 Dresden, Germany; matthias.krause@hzdr.de
3 Advanced Center for Aerospace Technologies (CATEC), C/Wilbur y Orville Wright 19, 41309 La Rinconada, Sevilla, Spain; iheras@catec.aero
4 C/Goles 26, 41002 Sevilla, Spain; grinonll@hotmail.com
5 Departamento de Física Aplicada I, Escuela Politécnica Superior, Universidad de Sevilla, Virgen de África 7, 41011 Sevilla, Spain
* Correspondence: rescobar1@us.es

Abstract: Titanium oxide films were deposited at room temperature and with no applied bias using a filtered cathodic vacuum arc (FCVA) system in a reactive oxygen environment. The dependence of film growth on two process parameters, the working pressure (*Pw*) and the O_2 partial pressure (p_{O2}), is described in detail. The composition, morphological features, crystalline structure, and optical properties of the deposited films were systematically studied by Rutherford Back Scattering (RBS), Scanning Electron Microscopy (SEM), X-Ray diffraction (XRD), Raman Spectroscopy, UV-vis spectroscopy, and spectroscopic ellipsometry. This systematic investigation allowed the identification of three different groups or growth regimes according to the stoichiometry and the phase structure of the titanium oxide films. RBS analysis revealed that a wide range of TiO_x stoichiometries ($0.6 < x < 2.2$) were obtained, including oxygen-deficient, stoichiometric TiO_2 and oxygen-rich films. TiO, Ti_2O_3, rutile-type TiO_2, and amorphous TiO_2 phase structures could be achieved, as confirmed both by Raman and XRD. Therefore, the results showed a highly versatile approach, in which different titanium oxide stoichiometries and crystalline phases especially suited for diverse optical applications can be obtained by changing only two process parameters, in a process at room temperature and without applied bias. Of particular interest are crystalline rutile films with high density to be used in ultra-high reflectance metal-dielectric multilayered mirrors, and reduced-TiO_2 rutile samples with absorption in the visible range as a very promising photocatalyst material.

Keywords: titanium oxide films; filtered cathodic vacuum arc; rutile; optical coatings

Citation: Guillén, E.; Krause, M.; Heras, I.; Rincón-Llorente, G.; Escobar-Galindo, R. Tailoring Crystalline Structure of Titanium Oxide Films for Optical Applications Using Non-Biased Filtered Cathodic Vacuum Arc Deposition at Room Temperature. *Coatings* **2021**, *11*, 233. https://doi.org/10.3390/coatings11020233

Academic Editor: Christian Mitterer

Received: 31 December 2020
Accepted: 10 February 2021
Published: 15 February 2021

1. Introduction

Great attention has been focused on the different polymorphs and stoichiometries of titanium oxide thin films due to their application in a wide range of fields [1]. In the case of TiO_2 films, different crystalline phases are described, namely anatase, rutile, and brookite, which determine its properties and applications. The first two phases are the most used for technological applications. TiO_2 films with anatase phase are used as self-cleaning windows, antifogging glass, self-sterilizing, and anti-bacterial tiles, as well as for water purification devices [2]. Anatase TiO_2 is also known for its use as a photoanode in Dye Sensitized Solar Cells [3]. Among the TiO_2 phases, the rutile phase is chemically and thermodynamically more stable and has the highest refractive index and hardness [4]. Some technological applications of rutile films include its use as optical coatings, as a dielectric layer in microelectronic applications, and as a protective layer on biomedical implants [5]. Rutile-type TiO_2 films can be especially suited as the high refractive index material of choice in combination with a low refractive index material as SiO_2 to achieve highly reflecting multilayer stacks in dielectric mirrors [6]. Regarding amorphous TiO_2, it has been synthesized as a tinted or enhanced photocatalyst, used to purify dye-polluted water, and applied to resistive random access memory applications [7].

In addition to the possibility to fabricate different crystalline phases, the desired properties of the TiO_2 films can be tuned by slight changes in their stoichiometry. TiO_2 is a wide bandgap semiconductor (3.0 and 3.2 eV for rutile and anatase, respectively), which is photoactive when irradiated with UV light. To widen the use of TiO_2 as a photocatalyst, its activity needs to be extended beyond the ultraviolet regime into the visible region [8]. In oxygen-deficient TiO_x films ($x < 2$), oxygen vacancy energy states are formed below the conduction band minimum [8]. This results in a reduction in the energy required for photoexcitation of electrons, making it possible for the films to absorb visible light. Oxygen-deficient TiO_2 films also have potential application as electrode material for alkaline-electrolyte batteries since they can be electrically conductive and yet maintain their corrosion-resistant properties [9]. However, there is no industrially scalable process available for high throughput synthesis of oxygen-deficient TiO_2.

The final properties of titanium oxide films strongly depend on the deposition technique employed for their fabrication (i.e., sol-gel, Chemical Vapour Deposition (CVD), Physical Vapour Deposition (PVD)). In particular, TiO_2 films deposited at low temperatures by most deposition methods exhibit an amorphous phase. Film crystallization is usually performed either by substrate heating during deposition or by a post-deposition annealing process. In addition, in PVD processes, very often the application of a bias to the substrate is required to obtain TiO_2 crystalline films. However, a myriad of recent applications require the TiO_2 films to be coated on heat-sensitive and/or non-conductive substrates (e.g., flexible electronics, antireflective coatings on plastic lenses, etc.). Therefore, it is important to optimize growth conditions in order to obtain crystalline films and/or the desired stoichiometry at low temperature using any type of substrate. To achieve that, it would be interesting to rely on other parameters of the deposition process different from the applied bias or the temperature. In this regard, there are some previous studies on the effect of the working pressure and the O_2 partial pressure on titanium oxide film formation [4,10].

Among the different PVD techniques, filtered cathodic vacuum arc (FCVA) has the highest plasma ionization ratio. This results in denser thin films and better adhesion than those obtained by technologies like magnetron sputtering or thermal evaporation [11]. Hence, this technique is especially suited for optical applications. The substrate temperature and the ion energy mainly determine phase formation during deposition by FCVA. In the case of TiO_2, amorphous films are obtained at low temperature and low ion energy. Therefore, synthesizing crystalline TiO_2 by FCVA usually requires either heating or biasing the substrate during deposition [5,12]. The phase diagram proposed by Löbl et al. [13] shows that with particle energies between 4–30 eV, mixed phases of rutile and anatase can be obtained at room temperature, meanwhile only at particle energies over 30 eV pure rutile phases can be obtained. These high ion energies are in the range of a FCVA deposition process.

In this work, a systematic study of the effect on titanium oxide films of the working pressure and oxygen partial pressure in a FCVA deposition process is carried out. We demonstrate that a fine tune of these parameters leads to the formation of a wide variety of stoichiometries and crystalline phases, from amorphous to rutile, at room temperature without the application of any bias during the deposition.

2. Materials and Methods

Titanium oxide films were grown on Si (100) and on glass substrates by means of a DC Filtered Cathodic Vacuum Arc (FCVA) system PFCVA-450 from Plasma Technology Limited, Hong Kong. The system is provided by a 90° curved electromagnetic filter which guides the ions in the plasma to the deposition chamber. The substrates were placed

240 mm away from the filter duct exit. For each deposition process, base pressure in the chamber was ~2 × 10^{-3} Pa. Prior to deposition, samples were sputter cleaned by Ar^+ bombardment for 15 min. In a FCVA system like the one used in this work, when no bias is applied during the process, the temperature at the surface of the sample reaches a maximum of ~100 °C after 40 min of deposition process. A Ti cathode with 65 mm diameter was used for the deposition of the films in an oxygen/argon atmosphere. The total working gas flow was fixed at 60 sccm, and different percentages of reactive gas during depositions were obtained by changing the Ar/O_2 ratio. The cathode current used was 50 A. All the depositions processes were carried out at room temperature. Deposition time was 15 min and the substrate holder was rotating during the whole process at 5 rpm. Depth-resolved composition of the samples was determined by Rutherford backscattering spectroscopy (RBS) using $^{4+}$He incident ions with an energy beam of 1.7 MeV. The data were acquired with a silicon barrier detector located at a backscattering angle of 170°, with an energy detector resolution of 13 keV. The experimental spectra were fitted using the SIMNRA software [14]. The thickness of the films was measured using a mechanical stylus profilometer (Veeco Dektak 150). Morphology was analyzed by scanning electron microscopy (SEM) with a Hitachi S4800 SEM-FEG microscope of high resolution (1–3 nm), equipped with a Bruker X flash 4010 EDX detector with a resolution of 133 eV, field emission gun, and STEM detector system. The phase structure of the thin films was determined by X-ray diffraction employing grazing incidence geometry (GIXRD) using a Rigaku Ultima IV diffractometer with Cu-Kα radiation (λ = 1.5406 Å). The incident angle was 0.4°, and the XRD patterns were measured in the diffraction angle range of 20–100° in steps of 0.02°. Raman spectra were measured in 180° backscattering geometry using a micro Raman LabramHR system (Horiba GmbH) equipped with a LN2 cooled CCD detector. For excitation, the frequency-doubled 532 nm emission of a Nd:YAG laser was focused to a 1 μm spot with 1 mW power on the sample surface. Spectroscopic Ellipsometry (SE) was used to determine the refractive index n of the samples. Measurements were performed with a rotating compensator ellipsometer M-2000FI (J.A. Woollam, Inc.) in the wavelength range of 211–1688 nm with a fixed angle of polarized light incidence and reflection of 75°.

3. Results

3.1. Thin Film Growth and Film Composition

The deposition of the films was performed at three different working pressures (P_w), 0.1, 0.25, and 0.5 Pa. For each P_w, the oxygen/argon ratio was varied by changing the O_2 gas flow from ~20 to ~80%. Consequently, the oxygen partial pressure (p_{O2}) during deposition ranged from 0.02 to 0.41 Pa. The O/Ti ratio for the different films obtained by RBS are shown in Figure 1. RBS profiles (not shown) indicated that the composition is uniform through the thickness of the films. It was found that the samples could be divided into three main groups according to their stoichiometry. In Figure 1, it is shown that the group 1 (G1) samples have an O/Ti ratio lower than 2 ($0.6 \leq$ O/Ti ≤ 1.6), samples in group 2 (G2) exhibit an O/Ti close to 2 ($1.9 \leq$ O/Ti ≤ 2.0), and finally, samples belonging to group 3 (G3) have an O/Ti ratio higher than 2 (O/Ti ≥ 2.1). This classification of the deposited films in three different groups is further supported by the differences in crystallinity of the samples, as will be shown by XRD and Raman results in Section 3.2. G1 samples are obtained at the lowest working pressure (0.1 Pa) or at the intermediate pressure (0.25 Pa) and low oxygen partial pressure (<0.1 Pa). Stoichiometric TiO_2 samples in G2 can be obtained at 0.25 Pa working pressure provided that the p_{O2} is over a threshold value of~> 0.1 Pa. Finally, over stoichiometric (G3) films are obtained for the working pressure of 0.5 Pa in the entire oxygen partial pressure range analyzed.

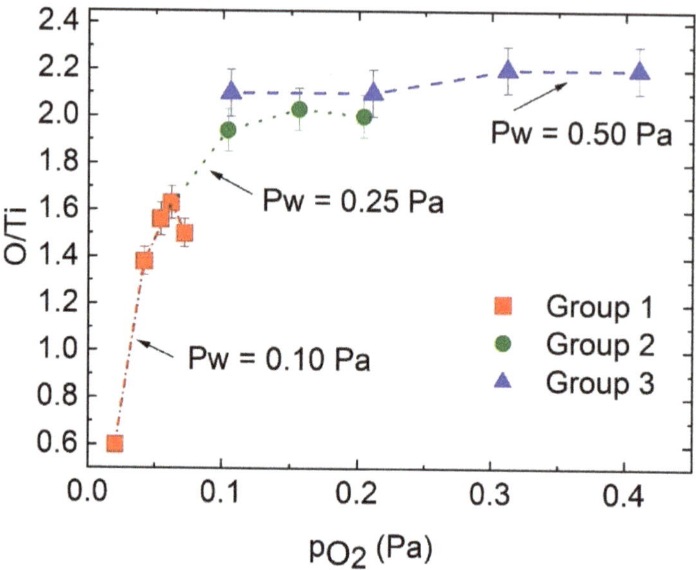

Figure 1. O/Ti ratio obtained from RBS for the different samples as a function of p_{O2}. The different groups are indicated by color and legend code: group 1 (square, red), group 2 (circle, green) and group 3 (triangle, blue). Samples deposited at the same P_w are connected through lines, and the value of the working pressure is indicated in the graph.

In Figure 2, a summary of the different deposited samples is depicted, indicating the three identified groups. All the samples in G1 are opaque. With respect to G2, visual inspection revealed that a slight variation in the stoichiometry of the samples (from O/Ti = 1.9 to O/Ti = 2) led to strong variation in transmittance. This can be observed by comparing G2 samples prepared at % O_2 gas flow of 40% and 60%. For the former, a reduced TiO_2 material is obtained (O/Ti = 1.9). TiO_2 bulk reduction results in color centers, which produce a pronounced color change of TiO_2 single crystals from initially transparent to visible light to eventually exhibiting a dark blue colour [2]. A similar result is observed in the current work for the reduced TiO_2 sample, which exhibits a blueish color. The stoichiometric TiO_2 samples prepared at 60 and 80% O_2 gas flow are transparent, as expected for a wide band gap semiconductor. Finally, all G3 samples are transparent, regardless of the oxygen partial pressure used during the deposition process.

In Figure 3, the thicknesses of the different samples obtained by profilometry are plotted as a function of the oxygen partial pressure, indicating also the samples that were deposited at the same working pressure. A gradient color can be observed in transparent samples due to interferences (Figure 2), which must be caused by non-homogeneity of the thickness along the substrate. For this reason, at least three measures of the thickness in different substrate positions were performed and the average and error values were obtained.

It is observed that the sample thickness is affected by both, the p_{O2} and the P_w. Higher thicknesses are obtained for lower working pressures and/or lower oxygen partial pressures. However, the working pressure is the factor that affects final thickness the most. For similar p_{O2} around 0.2 Pa, a sample deposited at Pw 0.25 Pa exhibits ~700 nm thickness, and one deposited at Pw 0.50 Pa is only 300 nm thick. At similar oxygen partial pressure, thickness decreases with working pressure, as the main free path of the species diminishes. Deposition rates were ~56 nm/min for G1, 45 nm/min for G2, and 24 nm/min for G3.

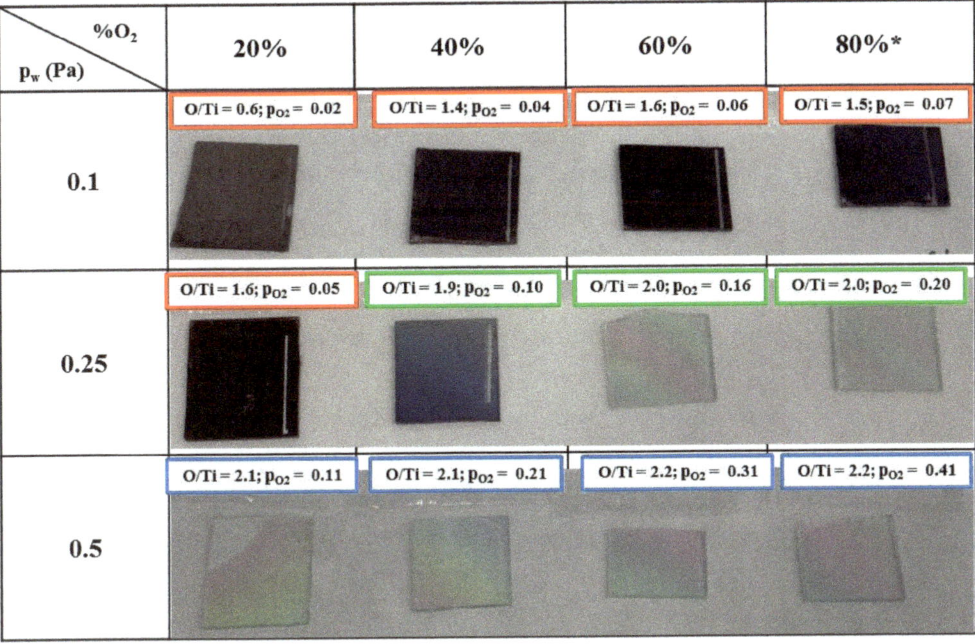

P_w (Pa) \ %O_2	20%	40%	60%	80%*
0.1	O/Ti = 0.6; p_{O2} = 0.02	O/Ti = 1.4; p_{O2} = 0.04	O/Ti = 1.6; p_{O2} = 0.06	O/Ti = 1.5; p_{O2} = 0.07
0.25	O/Ti = 1.6; p_{O2} = 0.05	O/Ti = 1.9; p_{O2} = 0.10	O/Ti = 2.0; p_{O2} = 0.16	O/Ti = 2.0; p_{O2} = 0.20
0.5	O/Ti = 2.1; p_{O2} = 0.11	O/Ti = 2.1; p_{O2} = 0.21	O/Ti = 2.2; p_{O2} = 0.31	O/Ti = 2.2; p_{O2} = 0.41

Figure 2. Summary table of deposited samples with varying % of Ar/O_2 gas flow, and P_w including: picture of each sample deposited on glass, p_{O2} during deposition and elemental at.% (O/Ti ratio) obtained by RBS. The code color indicates the samples belonging to group 1 (red), 2 (green), and 3 (blue). * For the lower working pressure (0.1 Pa), it was not possible to perform the deposition process with more than 70% of oxygen gas flow.

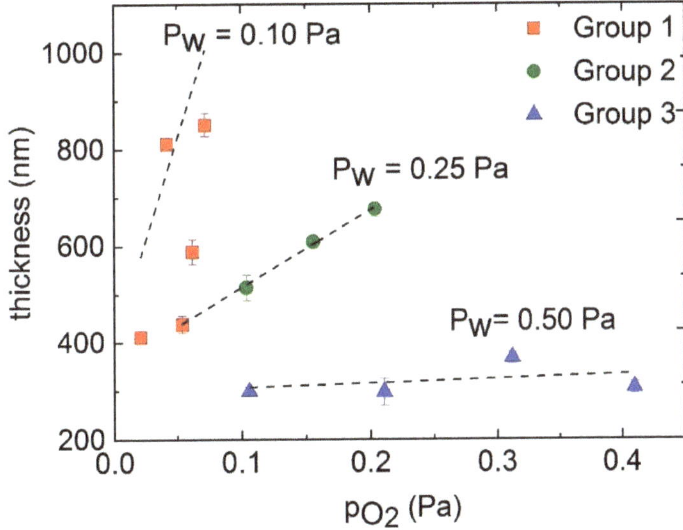

Figure 3. Coating thickness on silicon as a function of p_{O2} for samples belonging to group 1 (square, red), 2 (circle, green), and 3 (triangle, blue). A linear fit has been obtained for the samples deposited with the same working pressure. The total deposition time was 15 min in all cases.

Provided the film thickness is known from profilometry, the density of TiO_2 films can be obtained from RBS data [15]. Samples in G2 exhibit a density ~4.5 g cm^{-3}, meanwhile the density of samples in G3 is ~3.6 g cm^{-3}. As will be shown later, these results are in line with the crystallinity of the samples. Bulk theoretical values are 4.23 g cm^{-3} for rutile and 3.6 g cm^{-3} for amorphous TiO_2 [16]. Bendavid et al. [17] reported a density of 3.62 g cm^{-3} for TiO_2 amorphous films deposited by FCVA.

3.2. Structure and Morphology

Cross-sectional SEM analysis of the morphology revealed remarkable differences among the three groups. A columnar structure is observed for G1. High energy particle bombardment at low pressure can be the cause for the elongated faceted texture of the film (Figure 4a,b). For G2, two growth regimes are observed in the SEM images. During the first stages of the deposition process, the samples grow without any type of texturized structure (up to approximately 300 nm). Subsequently, the films develop a clear columnar structure (Figure 4c,d). This is also in line with the crystallinity observed in G2 samples, as confirmed by XRD experiments. Less energetic particle bombardment at higher pressure resulted in featureless and non-textured films, which is an indication of amorphous nature in G3 samples (Figure 4e,f).

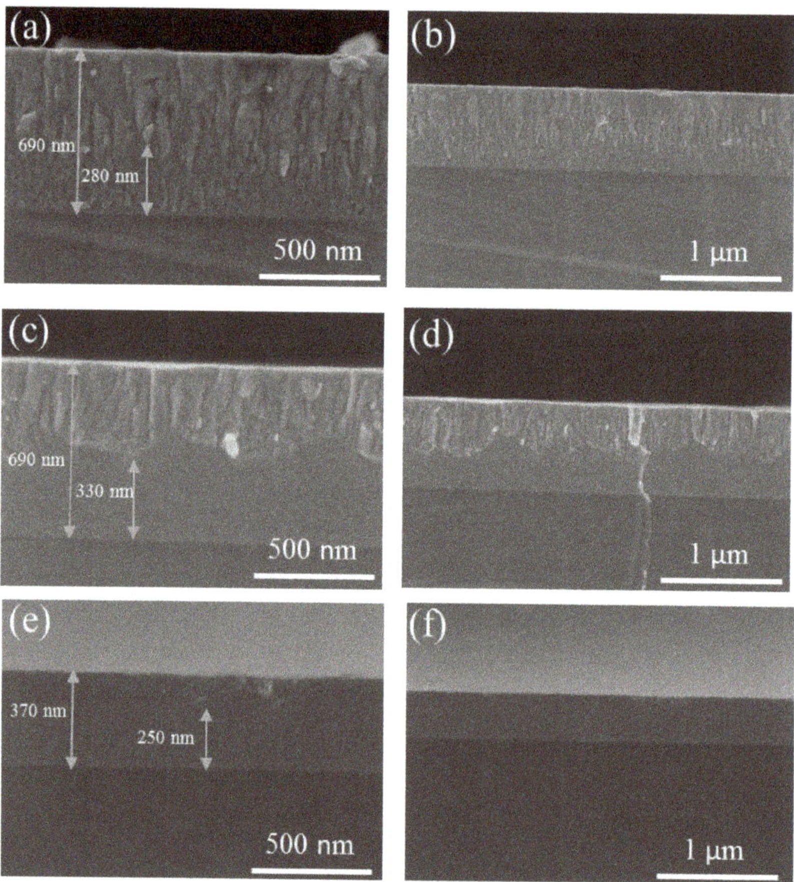

Figure 4. Representative SEM cross-section images for samples for G1 (**a**,**b**), sample deposited at *Pw* 0.1 Pa, p_{O2} = 0.06 Pa; G2 (**c**,**d**), sample deposited at *Pw* 0.25 Pa, p_{O2} = 0.20 Pa; and G3 (**e**,**f**), at Pw 0.5 Pa, p_{O2} = 0.31 Pa.

The Glancing Incidence XRD (GIXRD) patterns of selected films for the different groups are represented in Figures 5 and 6. For the crystalline samples, interplanar spacing (*d*) and lattice constants (*a* and *c*) were calculated as an average of the main observed peaks. The crystallite sizes obtained from Scherrer's formula [18] are also calculated for the TiO_2 stoichiometric samples. The results are summarized in Table 1.

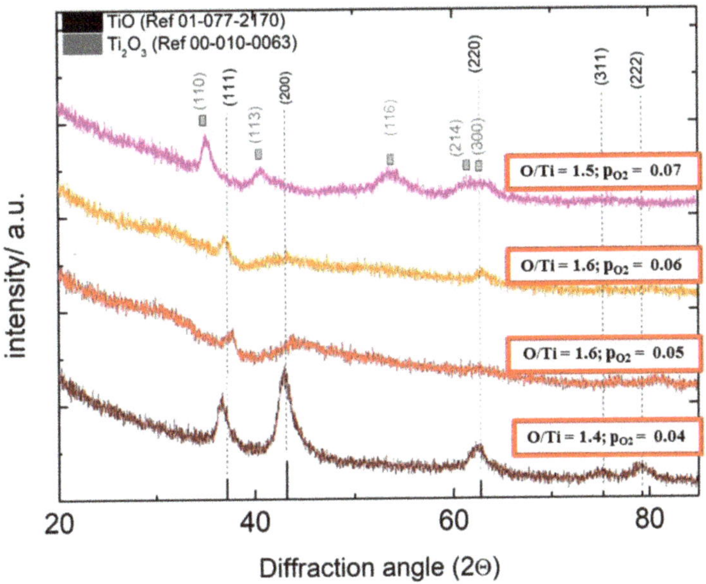

Figure 5. GIXRD patterns for samples of G1 deposited on silicon substrates. TiO and Ti_2O_3 reference peaks are included in the graph.

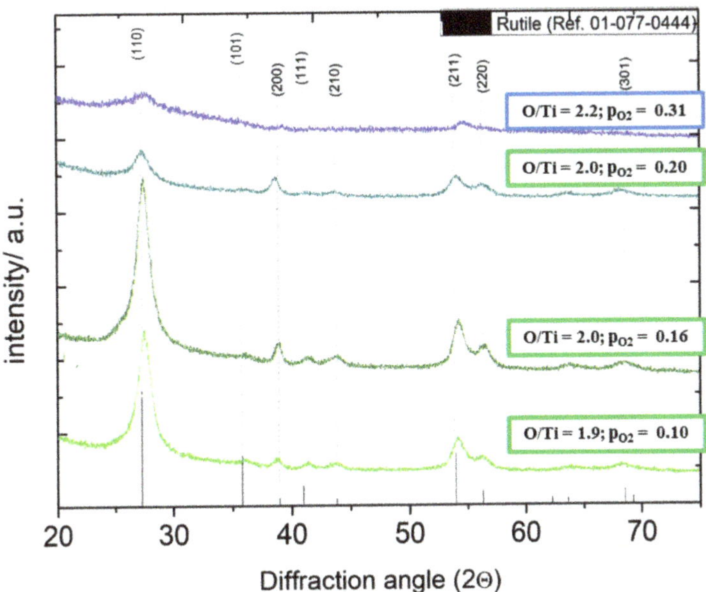

Figure 6. Glancing incidence XRD patterns for samples of G2, and a representative sample of G3. Rutile-type TiO_2 reference peaks are included in the graph.

Table 1. Peak position, crystal planes, interplanar spacing, and lattice constants for TiO and Ti_2O_3 reference cards and samples in group 1.

p_{O2} (Pa)	Peak Position (2θ)	(h k l)	d (Å)	a (Å)	c (Å)
Reference (TiO) (ICC card 01-077-2170)	37.18	(111)	2.416	4.18	4.18
	43.20	(200)	2.092		
	62.74	(220)	1.479		
	75.24	(311)	1.261		
0.04	36.60	(111)	2.45	4.25	4.25
	42.87	(200)	2.10	4.21	4.21
	62.38	(220)	1.48	4.21	4.21
	75.01	(311)	1.26	4.19	4.19
0.05	37.71	(111)	2.38	4.13	4.13
	43.72	(200)	2.07	4.14	4.14
	-	(220)	-	-	-
	76.54	(311)	1.24	4.12	4.12
0.06	37.01	(111)	2.42	4.20	
	-	(200)	-	-	
	62.92	(220)	1.47	4.17	
	-	(311)	-	-	
Reference (Ti_2O_3) (ICC card 00-010-0063)	34.85	(110)	2.57	5.13	13.65
	40.26	(113)	2.23		
	53.75	(116)	1.70		
	61.34	(214)	1.51		
	62.58	(300)	1.48		
	86.98	(226)	1.11		
0.07	35.00	(110)	2.53	-	-
	40.66	(113)	2.21		
	54.21	(116)	1.69		
	61.23	(214)	1.51		
	63.40	(300)	1.46		
	87.69	(226)	1.11		

In Figure 5, the diffractograms for the samples belonging to G1 are plotted. No peaks related to any TiO_2 crystalline phases are observed. For samples with p_{O2} of 0.04, 0.05, and 0.06 Pa, despite the different stoichiometries, the peaks can be attributed to TiO (ICC card 01-077-2170) with a cubic structure, (rock-salt space group 225, *Fm-3m*) in all the samples. Reference crystal peaks at 2θ = 36.60°, 42.87°, 62.38°, and 75.01° are associated with the crystal planes (111), (200), (220), and (311), respectively. For the sample with the strongest diffraction peaks in G1 (p_{O2} = 0.04), the diffraction peaks appear at slightly smaller diffraction angles (between 0.2 and 0.5°), corresponding to a larger lattice constant of the rocksalt-type sample than in the reference data set of TiO. The sample deposited at p_{O2} = 0.07 (O/Ti ratio of 1.5) shows diffraction peaks at 2θ = 35.00°, 40.66° and 54.21°. These peaks can be attributed to Ti_2O_3 (ICC card 00-010-0063), space group 167, *R-3c*, corresponding to the crystal planes (110), (113), and (116), respectively. Around 2θ = 62° a broad peak is observed which can be related to the contribution of crystal planes (214) and (300), related to peaks at 61.34° and 62.58° in the reference card (see Table 1). In general, the peaks are very broad, except for the one at 2θ = 35.00°. Parker et al [19] reported on the formation of the Ti_2O_3 layer by bombardment with Kr ions on TiO_2. A high dose ion impact led to preferential oxygen sputtering and/or internal precipitation of oxygen. Ion energy is higher in cathodic arc vacuum deposition, so, in the current work, the Ti_2O_3 phase could be formed at low working pressures (where high ion energy of the ions is expected) and with a high oxygen partial pressure.

A completely different scenario is observed for the three samples belonging to group 2 (Figure 6 and Table 2). The diffractograms of samples belonging to this group showed a pattern corresponding to rutile TiO_2 with tetragonal structure (space group 136, *P42-mnm*). The main rutile peak (110) is observed at 2θ = 27.39°, 27.38° and 27.3° for samples with p_{O2} of 0.10, 0.16, and 0.20 Pa, respectively. Hence these peaks are slightly shifted to higher angles with respect to the reference pattern (2θ = 27.29°) for this crystal plane (110). The other main peaks are observed at 2θ~54° and 56°, related to crystal planes (211) and (220). These three main peaks have been selected to calculate the lattice parameter (a and c) and the crystallite size. Other minor peaks confirming the rutile phase observed at 2θ~38.97, 40.97, 43.80, and 68.56° are associated to the crystal planes (200), (111), (210), and (301), respectively, according to the reference card.

Table 2. Peak position, crystal planes, interplanar spacing, and lattice constants for TiO_2 rutile reference cards and for samples in group 2.

p_{O2} (Pa)	Peak Position (2θ)	(h k l)	d (Å)	a (Å)	c (Å)	Crystallite Size (nm) Average
Reference Rutile TiO_2 (ICC card 01-077-0444)	27.29 53.98 56.30	(110) (211) (220)	3.265 1.697 1.632	4.61	2.97	-
0.10	27.39 54.06 56.07	(110) (211) (220)	3.252 1.694 1.639	4.60	2.99	6.60 ± 0.80
0.16	27.38 54.21 56.57	(110) (211) (220)	3.254 1.690 1.625	4.60	2.96	5.55 ± 0.99
0.20	27.33 54.04 56.33	(110) (211) (220)	3.254 1.690 1.625	4.61	2.98	6.00 ± 0.12

Finally, all group 3 samples showed only few broad and weak diffraction peaks, pointing to a predominantly X-ray amorphous phase structure with few nanocrystalline regions. See for example the diffractogram in Figure 6 corresponding to samples deposited at 0.31 Pa of p_{O2}. This phase structure can be related to the higher working pressure, and therefore to the lower energy of species impinging on the substrate. Only a small and very wide peak can be inferred at angles corresponding to the crystal plane (110) of the rutile phase. The rutile (110) surface is the most stable crystal face [2]. In the case of vapor deposition, the different crystallographic planes grow at different rates depending on their surface formation energies. The slight crystallization observed for films in G3 in the rutile phase for the plane (110) can be attributed to the lower surface formation energy of this plane [4]. In relation to the SEM results, this can be an indication of a change in the growth regime that could lead to crystalline structures.

Raman spectroscopy confirmed and complemented the data obtained by X-ray diffraction. For G1 samples grown at the oxygen partial pressures of 0.02 Pa $\leq p_{O2} <$ 0.06 Pa, the Raman spectra exhibit two broad band-like features centered at around 250 cm^{-1} and at around 550 cm^{-1} (Figure 7). The corresponding line widths of 200 to 300 cm^{-1} are typical for phonon density of states-like Raman spectra of amorphous materials. It is tempting to attribute the low-energy band (50 to 350 cm^{-1}) to titanium-based phonon states, and the high-energy band (400 to 700 cm^{-1}) to oxygen phonon-based states. Thus, the Raman spectra indicate a significant amount of amorphous structures, in addition to the fraction of rock salt-type TiO detected by XRD. Since crystalline TiO is Raman-silent, the combination of XRD and Raman data point to the coexistence of nanocrystalline and amorphous TiO in these four samples of G1. The Raman spectrum of the sample grown at an oxygen pressure of 0.07 Pa shows, in addition to broad lines at the same positions as those of the other four G1 samples, several narrower features (Figure 7). The most prominent ones among them, located at 263 cm^{-1} and 334 cm^{-1}, can be assigned to the most prominent lines reported for Ti_2O_3 at 269 cm^{-1} and 347 cm^{-1} (vertical bars in Figure 7) [20]. The observation of these lines supports the finding of Ti_2O_3 by XRD. Moreover, the Raman spectra reveal a significant amorphous phase fraction within the film grown at an oxygen pressure of 0.07 Pa at a working pressure of 0.10 Pa.

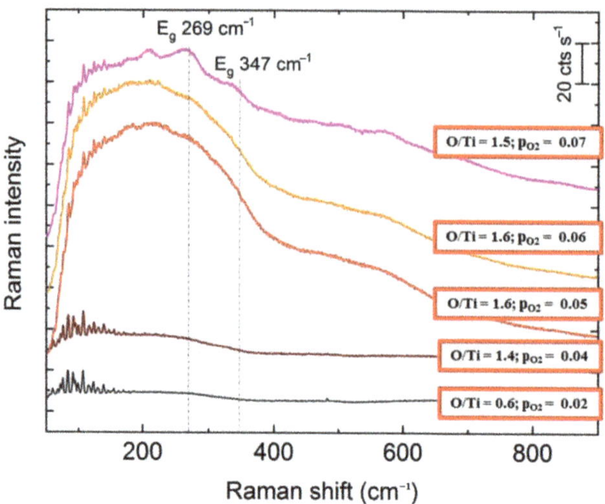

Figure 7. Raman spectra of samples in G1. Ti_2O_3 lines from Ref. [20] are indicated in the graph. The spectra were offset vertically for better visibility.

The Raman spectra of the films grown at a working pressure of 0.25 Pa with an oxygen partial pressure of 0.10 to 0.20 Pa (G2 samples) are dominated by strong lines at 450 ± 2 cm^{-1} and 608 ± 3 cm^{-1} (Figure 8). These lines point to the presence of nanocrystalline rutile-type TiO_2, whose strongest Raman lines are the E_g line at 447 cm^{-1} and the A_{1g} line at 612 cm^{-1} [21]. The line positions found for the films grown in this work are close to the literature values, shifted by +3 (E_g) and by −4 cm^{-1} (A_{1g}), respectively. The observed downshift of the A_{1g} line might be caused by the small size of the rutile-type TiO_2 crystals grown here. According to Swamy et al. a shift of −4 cm^{-1} would indicate a crystallite size of 10 nm [22]. This value is in very good agreement with that of 6 nm obtained by XRD for these samples (Table 2).

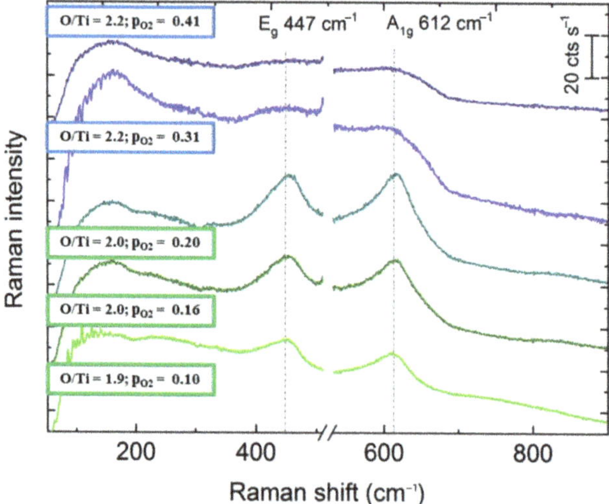

Figure 8. Raman spectra of samples in G2 and G3. Reference lines for the Eg and A_{1g} positions of rutile-type TiO_2 are included in the graph. Spectra were offset vertically for better visibility. The contribution of Si at 20 cm^{-1} was removed from the graph.

The Raman spectra of films belonging to group 3 exhibit two broad Raman lines at around 460 cm^{-1} and 608 cm^{-1} that indicate the presence of nanocrystalline Rutile-type TiO_2 in coexistence with the amorphous phase (Figure 8). According to Parker et al., the E_g line frequency scales linearly with the oxygen content in rutile-type TiO_2 [23]. Hence, the observed upshift of the E_g-derived line to 460 cm^{-1} in the G3 samples, compared to the E_g at 447 cm^{-1} in single crystals, can be qualitatively explained by the oxygen excess (O/Ti = 2.2) detected by RBS. The origin of the A_{1g} line downshift by −4 cm^{-1} compared to the reference value is presumably the same as for the G2 samples.

3.3. Optical Properties

Spectral transmittance and ellipsometry measurements were performed to correlate the changes observed in the stoichiometry and crystallinity of the films with their optical properties. Figure 9 shows the transmittance spectra of the films deposited on glass substrates for samples of the three different groups. Only stoichiometric samples or samples with O/Ti $\geq$ 2 exhibit a high transmittance in the visible and near IR region, in agreement with the images of the samples depicted in Figure 2. The transmission spectrum of G1 (O/Ti = 1.6) consists of a straight line in the whole UV-Vis-NIR. It is therefore a broad band absorber, presumably based on different color centers from differently charged ions (Ti^{2+}, Ti^{3+} (Ti^{+})) and maybe oxygen vacancies.

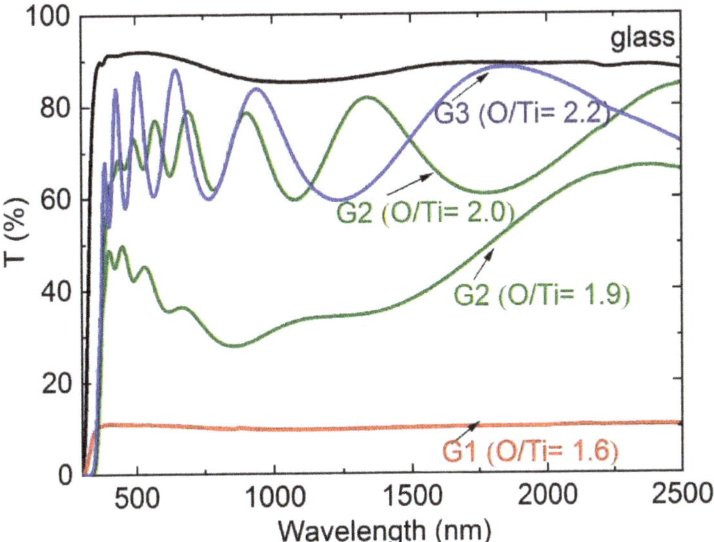

Figure 9. Transmittance of selected representative samples deposited on glass. The reference was air and the spectrum of glass substrate is also included in the graph. The code color indicates the samples belonging to group 1 (red), 2 (green), and 3 (blue), and the O/Ti ratio of the different samples is indicated next to each line.

As mentioned above, the transmittance of TiO_2 thin films is very sensitive to slight variation in the stoichiometry. The G2 sample deposited at p_{O2} = 0.1 Pa can be considered a reduced rutile sample. Breckenridge and Hosler [24] reported light absorption in the visible region in the case of slightly reduced oxygen-deficient TiO_2. Reduced rutile could imply some fraction of Ti^{3+}, or two O^{2-} vacancies for one Ti^{4+}. Several authors have reported the fabrication of reduced TiO_2 films [8,25,26].

In Figure 10, the spectral refractive indexes (a) and extinction coefficients (b) of representative samples for each group obtained from ellipsometry measurements are shown. Several groups have studied the optical constants of titanium oxide thin films under various deposition conditions. Refractive index and extinction coefficient both tended to decrease as the working pressure increases [9]. In contrast, in this work, there is no direct relation with the total working pressure, as the refractive index is highly dependent on the crystalline phase of the TiO_2. In general, amorphous films show a lower refractive index than that of crystalline film. Rutile structure exhibits the highest refractive index due to its superior density among all the TiO_2 phases. In this work, refractive indexes at a wavelength of 550 nm were 1.56, 2.63, and 2.57, measured for representative samples of groups 1, 2, and 3, respectively. The higher value of the refractive index obtained for group 2 is in line with the crystalline rutile structures observed in XRD and Raman. A refractive index of 2.57 at 550 nm (similar to the obtained for G3 samples) has been reported in the literature for amorphous TiO_2 films deposited at RT with no bias by FCVA [27]. This *n* is higher than the one obtained for amorphous films with other deposition techniques [9], due to the enhanced reactivity and high energy of the Ti ion beam impinging on the growing TiO_2 films by FCVA.

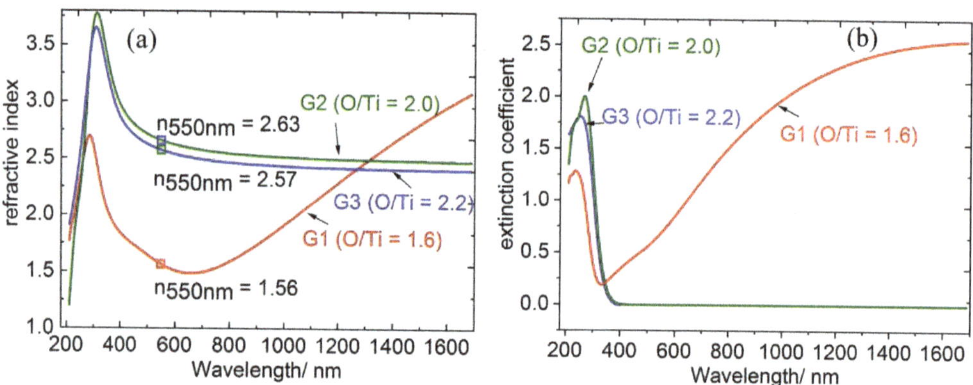

Figure 10. Refractive index (**a**) and extinction coefficient (**b**) in the 200–1700 nm wavelength range for one representative sample of group 1 (red), 2 (green), and 3 (blue). The refractive index at 550 nm is marked as an open square and the actual value is included in the graph.

4. Discussion

Titanium oxide films were grown by FCVA and the influence of the total working pressure and the oxygen partial pressure during the deposition process was systematically studied. A strong correlation between those parameters and the properties of the films was found, and three different growth regimes or groups were identified.

Samples belonging to group 1 were obtained (i) with the lowest working pressure used in this work (P_w = 0.1 Pa) and, (ii) with the intermediate pressure (P_w = 0.25 Pa) if the oxygen partial pressure was below 0.1 Pa. These samples exhibited a columnar structure in the SEM cross-section images and were opaque. Besides, all the samples shared an O/Ti ratio < 2, and the presence of crystalline phases different fromTiO_2. For samples with a p_{O2} of 0.04, 0.05, and 0.06 Pa, the combination of XRD and Raman pointed out the coexistence of nanocrystalline and amorphous TiO. For the samples deposited at p_{O2} = 0.7 Pa, both XRD and Raman revealed features, which were attributed to Ti_2O_3. The Raman spectra also revealed a significant amount of amorphous Ti_2O_3 phase in this film.

In the range corresponding to intermediate working pressures (0.25 Pa) and oxygen partial pressure above 0.10 Pa, we found the samples forming group 2. In addition to a stoichiometry close to TiO_2, they shared their crystalline structure. Both XRD and Raman indicated that the samples in this group exhibited a rutile TiO_2 structure. This crystalline structure provided the samples of this group with very interesting properties for optical applications. The rutile crystalline structure and the high film density (~4.5 g cm^{-3}) led to a higher refractive index (n_{550nm} = 2.62) in comparison to amorphous TiO_2 or rutile TiO_2 fabricated with less energetic deposition techniques and obtained by post-annealing processes. Samples of this group are very promising, for example, as high refractive index dielectric material in ultra-high reflectance metal-dielectric multilayered mirrors [6]. The sample deposited with the lowest oxygen partial pressure (p_{O2} = 0.10 Pa) is worthy of note. It exhibited similar characteristics to the other samples belonging to this group, with a stoichiometry close to TiO_2 and a rutile phase structure. However, contrary to them, it is not transparent but exhibited a blueish colour and was absorbed in the visible region. This reduced TiO_2 sample is very interesting as it preserves the properties of crystalline rutile TiO_2, while, and at the same time is photoactive in the visible region, being a very promising photocatalyst material, among other applications.

Finally, samples belonging to group 3 were obtained with the highest working pressure (P_w = 0.5 Pa), regardless of the oxygen partial pressure. All the samples are transparent and exhibit oxygen excess with O/Ti ratios above 2.1. The combination of SEM images, the lower values for density and refractive index compared to group 2, together with XRD and Raman results, pointed out the amorphous phase structure of these samples. However, a tendency of growth regimes from amorphous to crystalline TiO_2 could be inferred from a small rutile peak in XRD and broad Raman lines for Rutile- type TiO_2.

A summary of properties reported for TiO_2 films deposited by FCVA in previous references is included in Table 3. A final line has been added to the table with the results obtained in the current work.

Table 3. Summary of previous references of TiO_2 films deposited using FCVA, including information about the applied bias, temperature and working pressure during deposition, the observed crystalline phase, and the refractive index. The results observed in the current work are included in the last row.

Ref.	Bias	T°	P_w (Pa)	Crystalline phase	n
Zhang and Liu, 1998 [28]	−400 V	300 °C	0.04, 0.2	rutile	
Zhang et al., 1998 [29]	0 to 400 V	300 °C	0, 2	rutile (bias); amorphous (0V)	
Bendavid et al., 1999 [30]	0 to −400V	RT		anatase (0 V); amorphous (−50 V); rutile (−100 to −400 V)	anatase: 2.62; rutile: 2.72
Martin et al., 1999 [31]				amorphous	amorphous: 2.45
Takikawa et al., 1999 [32]			0.1–2.0	amorphous	
Bendavid et al., 2000 [17]	0 to −400V	RT	0.35	anatase (0 V); amorphous (−50 V); rutile (−100 to −400 V)	amorphous: 2.56; anatase: 2.62; rutile: 2.72
Z. W. Zhao et al., 2004 [33]	-	-	-	amorphous	2.57 at 550 nm
Z. Zhao et al., 2004 [27]	-	RT	0.03	amorphous	2.56 at 550 nm
Huang et al., 2006 [34]	0, −200, −500V	450 °C		amorphous (0V); anatase, (−200 V); rutile (−500 V)	
Kleiman et al., 2006 [35]		200–400 °C		amorphous (200 °C); anatase (400 °C)	
Kleiman et al., 2007 [36]	no	RT-400 °C		amorphous (<300 °C); anatase (>300 °C)	
Zhang et al., 2007 [37]	no	RT		amorphous	2.51 at 550 nm
Zhao, 2007 [38]				amorphous	
Zhang et al., 2007 [37]	0 to −900V			amorphous	
Bendavid et al., 2008 [39]	−50 V to −300 V	RT	0.3	anatase and some rutilo	2.59 at 550 nm
Zhirkov et al., 2011 [12]	no			amorphous	-
Zhirkov et al., 2011 [12]	0, −1, −50, −70 V	500 °C, 300 °C, 150 °C, 200 °C, 15 °C	0.6, 0.01, 0.006	500 °C, 300 °C, no bias: rutile 150 °C (0, −10, −50, −70 V): only a broad (101) rutile peak RT: no bias: amorphous	-
Arias et al., 2012 [5]	no	RT-570 °C	2–5	RT: amorphous (RT); anatase/rutile (in SS, 360 °C); anatase+rutile (400–500 °C); rutile (560 °C)	-

Table 3. *Cont.*

Ref.	Bias	T°	P_w (Pa)	Crystalline phase	*n*
Paternoster et al., 2013 [40]		100 °C	P_{O2} = 0.06	amorphous (RT); rutile (Tsub ≥ 300 °C)	-
Aramwit et al., 2014 [3]	0 or −250 V	RT	0.013, 0.13, 1.3, 13	amorphous (0 V), some rutile (bias)	-
Franco Arias et al., 2017 [16]	−120	300 °C or 400 °C	5–8	small rutile and anatase peaks (no Ti interlayer); rutile (with Ti interlayer)	-
Current work	**0**	**RT**	**0.02–0.41**	**TiO/Ti_2O_3/Rutile TiO_2 and amorphous TiO_2**	**rutile: 2.63 amorphous: 2.57**

It is commonly observed in PVD that the phase formation is mainly determined by the temperature and ion energy [41]. At room temperature, without external heating and substrate bias, titanium oxide thin films deposited by FCVA are usually amorphous in structure [9]. Zhang et al. [29] observed that when the substrate bias was −100 V, films exhibited rutile-type structure with XRD (110), (101), and (111) diffraction peaks, the (111) peak being dominant. These authors further demonstrated that without bias, the film structure remained amorphous. This is a result generally observed, as shown in Table 3. Some exceptions are found, for example in the works by Bendavid et al. [17,30]. These authors reported an anatase structure without substrate bias at room temperature at an oxygen pressure of 0.35 Pa. When the substrate bias was applied from zero to −400 V, the film structure varied from anatase (0 V) to amorphous phase at −50 V, evolving into a rutile structure with (101) orientation dominating in the films when the applied voltage is <−100 V. These authors ascribed the formation of the rutile phase to two concurrent factors: (i) the high concentration of ionic species (e.g., Ti^+), a typical feature of plasma produced by FCVA, and (ii) the high energy of particles impinging on the substrate, due to the applied bias [17,30].

However, in the present work, XRD and Raman indicated that the rutile phase was obtained at room temperature without the need for any applied bias. Besides, no anatase phase formation was observed within the range of studied process parameters, although the rutile phase usually requires higher activation energy for its formation [42]. This is, to the best knowledge of the authors, the first time that the fabrication of rutile-type TiO_2 at low temperature with no applied bias is reported for a FCVA process. This paves the way for the deposition of this TiO_2 crystalline phase with outstanding optical properties in transparent substrates such as glass, which is non-conductive and also on heat-sensitive substrates.

In summary, this work describes a very versatile method for the deposition of titanium oxide films with different stoichiometries and crystalline phases, demonstrating the great potential of FCVA for the deposition of these films for optical applications.

Author Contributions: Conceptualization, E.G., M.K. and R.E.-G.; Data curation, E.G.; Formal analysis, E.G., M.K. and R.E.-G.; Investigation, E.G., I.H. and G.R.-L.; Methodology, R.E.-G.; Project administration, R.E.-G.; Supervision, R.E.-G.; Visualization, E.G.; Writing—original draft E.G., M.K. and R.E.-G.; Writing—review & editing, E.G., M.K., I.H., G.R.-L. and R.E.-G. All authors have read and agreed to the published version of the manuscript.

Funding: This project was partially supported by H2020 RISE project "Framework of Innovation for Engineering of New Durable Solar Surfaces (FRIENDS2, GA-645725)".

Institutional Review Board Statement: Not applicable.

Informed Consent Statement: Not applicable.

Data Availability Statement: Not applicable.

Acknowledgments: Support by the Ion Beam Centre (IBC, Helmholtz-Zentrum Dresden-Rossendorf) is gratefully acknowledged. The authors would like to thank Ma C. Jiménez de Haro (Instituto Ciencia y Tecnología de Materiales de Sevilla) for the SEM service. Any opinions, findings, and conclusions or recommendations expressed in this material are those of the authors and do not necessarily reflect those of the host institutions or funders.

Conflicts of Interest: The authors declare no conflict of interest.

References

1. Chen, X.; Mao, S.S. Titanium dioxide nanomaterials: Synthesis, properties, modifications, and applications. *Chem. Rev.* **2007**, *107*, 2891–2959. [CrossRef]
2. Diebold, U. The Surface Science of Titanium Dioxide. *Surf. Sci. Rep.* **2003**, *48*, 53–229. [CrossRef]
3. Aramwit, C.; Intarasiri, S.; Bootkul, D.; Tippawan, U.; Supsermpol, B.; Seanphinit, N.; Ruangkul, W.; Yu, L. Effects of Filtered Cathodic Vacuum Arc Deposition (FCVAD) Conditions on Photovoltaic TiO_2 Films. *Appl. Surf. Sci.* **2014**, *310*, 266–271. [CrossRef]
4. Pradhan, S.S.; Pradhan, S.; Bhavanasi, V.; Sahoo, S.; Sarangi, S.; Anwar, S.; Barhai, P. Low Temperature Stabilized Rutile Phase TiO_2 Films Grown by Sputtering. *Thin Solid Films* **2012**, *520*, 1809–1813. [CrossRef]
5. Arias, L.M.F.; Kleiman, A.; Heredia, E.; Márquez, A. Rutile Titanium Dioxide Films Deposited With a Vacuum Arc at Different Temperatures. *J. Phys. Conf. Ser.* **2012**, *370*, 012027. [CrossRef]
6. Jiménez-Solano, A.; Anaya, M.; Calvo, M.E.; Alcon-Camas, M.; Alcañiz, C.; Guillén, E.; Martínez, N.; Gallas, M.; Preussner, T.; Escobar-Galindo, R.; et al. Aperiodic Metal-Dielectric Multilayers As Highly Efficient Sunlight Reflectors. *Adv. Opt. Mater.* **2017**, *5*, 1600833. [CrossRef]
7. Prasai, B.; Cai, B.; Underwood, M.K.; Lewis, J.P.; Drabold, D.A. Properties of Amorphous and Crystalline Titanium Dioxide from First Principles. *J. Mater. Sci.* **2012**, *47*, 7515–7521. [CrossRef]
8. Dhumal, S.Y.; Daulton, T.L.; Jiang, J.; Khomami, B.; Biswas, P. Synthesis of visible light-active nanostructured TiOx ($x < 2$) photocatalysts in a flame aerosol reactor. *Appl. Catal. B Environ.* **2009**, *86*, 145–151.
9. Tay, B.; Zhao, Z.; Chua, D. Review of Metal Oxide Films Deposited by Filtered Cathodic Vacuum Arc Technique. *Mater. Sci. Eng. R Rep.* **2006**, *52*, 1–48. [CrossRef]
10. Lim, J.C.; Song, K.J.; Park, C. The Effect of Deposition Parameters on the Phase of TiO_2 Films Grown by RF Magnetron Sputtering. *J. Korean Phys. Soc.* **2014**, *65*, 1896–1902. [CrossRef]
11. Berkeley, L. Energetic Deposition Using Filtered Cathodic Arc Plasmas. *Vacuum* **2002**, *67*, 673–686.
12. Zhirkov, I.S.; Paternoster, C.; Delplancke-Ogletree, M.P. Titanium Oxide Thin Film Deposition by Pulsed Arc Vacuum Plasma. *J. Phys. Conf. Ser.* **2011**, *275*, 12019. [CrossRef]
13. Löbl, P.; Huppertz, M.; Mergel, D. Nucleation and Growth in TiO_2 Films Prepared by Sputtering and Evaporation. *Thin Solid Films* **1994**, *251*, 72–79. [CrossRef]
14. Mayer, M. *SIMNRA User's Guide*; Max-Plank-Institut für Plasmaphysuk: Garching, Germany, 1997.
15. Nastasi, M.; Mayer, J.W.; Wang, Y. *Ion Beam Analysis: Fundamentals and Applications*; CRC Press: Boca Raton, FL, USA, 2014.
16. Arias, L.F.; Kleiman, A.; Vega, D.; Fazio, M.; Halac, E.; Márquez, A. Enhancement of Rutile Phase Formation in TiO_2 Films Deposited on Stainless Steel Substrates With a Vacuum Arc. *Thin Solid Films* **2017**, *638*, 269–276. [CrossRef]
17. Bendavid, A.; Martin, P.J.; Takikawa, H. Deposition and Modication of Titanium Dioxide Thin Films by Filtered Arc Deposition. *Thin Solid Film.* **2000**, *360*, 241–249. [CrossRef]
18. Patterson, A.L. The Scherrer Formula for X-ray Particle Size Determination. *Phys. Rev.* **1939**, *56*, 978–982. [CrossRef]
19. Parker, T.; Kelly, R. Ion-Impact Chemistry in the System Titanium-Oxygen (studies on Bombardment-Enhanced conductivity—III). *J. Phys. Chem. Solids* **1975**, *36*, 377–385. [CrossRef]
20. Mooradian, A.; Raccah, P.M. Raman Study of the Semiconductor-Metal Transition in Ti_2O_3. *Phys. Rev. B* **1971**, *3*, 4253–4256. [CrossRef]
21. Porto, S.P.S.; Fleury, P.A.; Damen, T.C. Raman Spectra of TiO_2, MgF_2, ZnF_2, FeF_2, and MnF_2. *Phys. Rev.* **1967**, *154*, 522–526. [CrossRef]
22. Swamy, V.; Muddle, B.C.; Dai, Q. Size-Dependent Modifications of the Raman Spectrum of Rutile TiO_2. *Appl. Phys. Lett.* **2006**, *89*, 163118. [CrossRef]
23. Parker, J.C.; Siegel, R.W. Calibration of the Raman Spectrum to the Oxygen Stoichiometry of Nanophase TiO_2. *Appl. Phys. Lett.* **1990**, *57*, 943–945. [CrossRef]
24. Breckenridge, R.G.; Hosler, W.R. Electrical Properties of Titanium Dioxide Semiconductors. *Phys. Rev.* **1953**, *91*, 793–802. [CrossRef]
25. Cronemeyer, D.C. Infrared Absorption of Reduced Rutile TiO2Single Crystals. *Phys. Rev.* **1959**, *113*, 1222–1226. [CrossRef]
26. Ihara, T.; Miyoshi, M.; Ando, M.; Sugihara, S.; Iriyama, Y. Preparation of a Visible-Light-Active TiO_2 Photocatalyst by RF Plasma Treatment. *J. Mater. Sci.* **2001**, *36*, 4201–4207. [CrossRef]
27. Zhao, Z.; Tay, B.K.; Yu, G. Room-Temperature Deposition of Amorphous Titanium Dioxide Thin Film With High Refractive Index by a Filtered Cathodic Vacuum Arc Technique. *Appl. Opt.* **2004**, *43*, 1281–1285. [CrossRef] [PubMed]
28. Zhang, F.; Liu, X. Effect of O_2 Pressure on the Preferred Orientation of TiO_2 Films Prepared by Filtered Arc Deposition. *Thin Solid Films* **1998**, *326*, 171–174. [CrossRef]

29. Zhang, F.; Wang, X.; Li, C.; Wang, H.; Chen, L.; Liu, X. Rutile-Type Titanium Oxide Films Synthesized by Filtered Arc Deposition. *Surf. Coat. Technol.* **1998**, *110*, 136–139. [CrossRef]
30. Bendavid, A.; Martin, P.; Jamting, Å.; Takikawa, H. Structural and Optical Properties of Titanium Oxide Thin Films Deposited by Filtered Arc Deposition. *Thin Solid Films* **1999**, *355–356*, 6–11. [CrossRef]
31. Martin, P.; Bendavid, A.; Netterfield, R.; Kinder, T.; Jahan, F.; Smith, G. Plasma Deposition of Tribological and Optical Thin Film Materials With a Filtered Cathodic Arc Source. *Surf. Coat. Technol.* **1999**, *112*, 257–260. [CrossRef]
32. Takikawa, H.; Matsui, T.; Sakakibara, T.; Bendavid, A.; Martin, P.J. Properties of Titanium Oxide Film Prepared by Reactive Cathodic Vacuum Arc Deposition. *Thin Solid Films* **1999**, *348*, 145–151. [CrossRef]
33. Zhao, Z.; Tay, B.K.; Lau, S.P.; Yu, G. Optical Properties of Titania Films Prepared by off-Plane Filtered Cathodic Vacuum Arc. *J. Cryst. Growth* **2004**, *268*, 543–546. [CrossRef]
34. Huang, A.; Chu, P.K.; Wang, L.; Cheung, W.; Xu, J.; Wong, S. Fabrication of Rutile TiO_2 Thin Films by Low-Temperature, Bias-Assisted Cathodic Arc Deposition and Their Dielectric Properties. *J. Mater. Res.* **2006**, *21*, 844–850. [CrossRef]
35. Kleiman, A.; Márquez, A.; Lamas, D.G. Optimization of a DC Vacuum Arc to Obtain Anatase Phase TiO_2 Coatings. *Fourth Huntsville Gamma-Ray Burst Symp.* **2006**, *875*, 223–226. [CrossRef]
36. Kleiman, A.; Marquez, A.; Lamas, D. Anatase TiO_2 Films Obtained by Cathodic Arc Deposition. *Surf. Coat. Technol.* **2007**, *201*, 6358–6362. [CrossRef]
37. Zhang, M.; Lin, G.; Dong, C.; Wen, L. Amorphous TiO_2 Films With High Refractive Index Deposited by Pulsed Bias Arc Ion Plating. *Surf. Coat. Technol.* **2007**, *201*, 7252–7258. [CrossRef]
38. Zhao, S. Spectrally Selective Solar Absorbing Coatings Prepared by Dc Magnetron Sputtering. Ph.D. Dissertation, 2007. Available online: http://uu.diva-portal.org/smash/get/diva2:169700/FULLTEXT01.pdf (accessed on 31 December 2020).
39. Bendavid, A.; Martin, P.; Preston, E. The Effect of Pulsed Direct Current Substrate Bias on the Properties of Titanium Dioxide Thin Films Deposited by Filtered Cathodic Vacuum Arc Deposition. *Thin Solid Films* **2008**, *517*, 494–499. [CrossRef]
40. Paternoster, C.; Zhirkov, I.; Delplancke-Ogletree, M.-P. Structural and Mechanical Characterization of Nanostructured Titanium Oxide Thin Films Deposited by Filtered Cathodic Vacuum Arc. *Surf. Coat. Technol.* **2013**, *227*, 42–47. [CrossRef]
41. Gjevori, A.; Gerlach, J.; Manova, D.; Assmann, W.; Valcheva, E.; Mändl, S. Influence of Auxiliary Plasma Source and Ion Bombardment on Growth of TiO_2 Thin Films. *Surf. Coat. Technol.* **2011**, *205*, S232–S234. [CrossRef]
42. Šícha, J.; Musil, J.; Meissner, M.; Čerstvý, R. Nanostructure of Photocatalytic TiO_2 Films Sputtered at Temperatures below 200 °C. *Appl. Surf. Sci.* **2008**, *254*, 3793–3800. [CrossRef]

Article

Spectrally Selective Solar Absorber Coating of W/WAlSiN/SiON/SiO_2 with Enhanced Absorption through Gradation of Optical Constants: Validation by Simulation

K. Niranjan [1,2], Paruchuri Kondaiah [1], Arup Biswas [3], V. Praveen Kumar [1], G. Srinivas [1] and Harish C. Barshilia [1,2,*]

1 Nanomaterials Research Laboratory, Surface Engineering Division, CSIR-National Aerospace Laboratories, Bangalore 560017, India; niru.snakes@gmail.com (K.N.); paruchurikondaiah@gmail.com (P.K.); praveenkv@nal.res.in (V.P.K.); Sree9977@nal.res.in (G.S.)
2 Academy of Scientific and Innovative Research (AcSIR), Ghaziabad 201002, India
3 Atomic and Molecular Physics Division, Bhabha Atomic Research Centre, Mumbai 400085, India; arup.bi@gmail.com
* Correspondence: harish@nal.res.in

Abstract: The properties of spectrally selective solar absorber coatings can be fine-tuned by varying the thickness and composition of the individual layers. We have deposited individual layers of WAlSiN, SiON, and SiO_2 of thicknesses ~940, 445, and 400 nm, respectively, for measuring the refractive indices and extinction coefficients using spectroscopic ellipsometer measurements. Appropriate dispersion models were used for curve fitting of ψ and Δ for individual and multilayer stacks in obtaining the optical constants. The W/WAlSiN/SiON/SiO_2 solar absorber exhibits a high solar absorptance of 0.955 and low thermal emissivity of 0.10. The refractive indices and extinction coefficients of different layers in the multilayer stack decrease from the substrate to the top anti-reflection layer. The graded refractive index of the individual layers in the multilayer stack enhances the solar absorption. In the tandem absorber, WAlSiN is the main absorbing layer, whereas SiON and SiO_2 act as anti-reflection layers. A commercial simulation tool was used to generate the theoretical reflectance spectra using the optical constants are in well accordance with the experimental data. We have attempted to understand the gradation in refractive indices of the multilayer stack and the physics behind it by computational simulation method in explaining the achieved optical properties. In brief, the novelty of the present work is in designing the solar absorber coating based on computational simulation and ellipsometry measurements of individual layers and multilayer stack in achieving a high solar selectivity. The superior optical properties of W/WAlSiN/SiON/SiO_2 makes it a potential candidate for spectrally selective solar absorber coatings.

Keywords: spectrally selective absorber; multilayer stack; spectroscopic ellipsometry; optical constants; simulation

Citation: Niranjan, K.; Kondaiah, P.; Biswas, A.; Kumar, V.P.; Srinivas, G.; Barshilia, H.C. Spectrally Selective Solar Absorber Coating of W/WAlSiN/SiON/SiO_2 with Enhanced Absorption through Gradation of Optical Constants: Validation by Simulation. *Coatings* **2021**, *11*, 334. https://doi.org/10.3390/coatings11030334

Academic Editor: Alicia de Andrés

Received: 6 January 2021
Accepted: 11 March 2021
Published: 15 March 2021

1. Introduction

Solar energy is one of the abundantly available renewable sources and has drawn researchers' interest due to diminishing non-renewable energy (fossil fuels). The solar radiation is converted into thermal energy through photothermal conversion by means of concentrated solar power (CSP) plants, which is a suitable way of producing thermal energy and disposable electricity [1]. The solar absorber has a significant role in improving the overall efficiency of CSP, as solar absorber coating determines the photothermal conversion of incident solar radiation to heat energy [2]. A real solar absorber coating should possess high solar absorptance ($\alpha \geq 0.950$) in the solar spectrum range (0.25–2.5 μm) and very low thermal emissivity ($\varepsilon \leq 0.10$) in the infrared (IR) range (2.5–25 μm) [3]. Spectrally selective solar absorbers with high solar absorptance, low thermal emissivity and high thermal stability deposited using physical vapor deposited (PVD) are widely used in parabolic trough collectors operating at temperatures of 400 °C at maximum. In recent years, solar absorbers have drawn massive interest due to their novel optical properties in the solar spectral and IR range [4]. Solar selective coatings are broadly classified into five different types namely: (a) Intrinsic absorber, (b) semiconductor absorber, (c) textured surfaces, (d) multilayer stack, and (e) cermet-based absorbers. Amongst these, multilayer stack and cermet-based solar absorbers are widely investigated [5]. The multilayer stack and/or cermet-based solar absorber consists of an infrared reflector, an absorber and an anti-reflection layer. The metal interlayers such as W, Mo, Ni, and Ti in the multilayer stack are mainly used to reduce the thermal emissivity of the coating in the infrared region at high temperatures [6,7]. Tungsten (W) is a potential material as interlayer due to its excellent thermal stability, low infrared emissivity and also acts as diffusion barrier. In case of absorbers, a wide range of transition metal nitrides/oxynitride based solar absorber coatings have been developed due to the tuneable optical properties and high thermal stability at elevated temperatures. Transition metals like Cr, Mn, W, Ni, and Mo usually show good selectivity and good thermal stability via doping of nitrogen and oxygen to form respective nitrides, oxynitrides, and oxides as main absorber layers [8]. Various anti-reflection layers such as SiO_2, SiO_xN_y, Si_3N_4, TiO_2, Al_2O_3, and $AlSiO_y$ are used in solar absorbers based on the application [9–11]. The anti-reflection layer deposited on the absorber layer reduces the surface reflection and thereby reducing the reflection losses. In this regard, a multilayer stack is designed by optimizing the optical properties of individual layers to achieve high absorptance.

The optical properties of thin films are mainly dependent on the optical constants, i.e., refractive index (n) and extinction coefficient (k). The refractive index of the individual layers has a significant effect in designing a high absorptance solar selective coating. A gradation in refractive index from top anti-reflection layer to bottom of the substrate in the multilayers stack significantly increases the absorption of the coating [6]. A wide range of wavelengths in solar radiation can experience enhanced absorption due to multiple reflections at layer interfaces. Two factors are usually attributed to the absorption in thin films: One is because the phase difference between the top and bottom layers of the coating surface accounts for destructive interference of light and the other is band-to-band transitions. The refractive indices and extinction coefficients of the individual layers in the multi-layer stack determine the reflectance behavior, thereby, helping in better under-standing of the absorption mechanism and high solar selectivity (α/ε). In the past decade, several reports have been devoted on the effect of refractive index and extinction coefficient on the optical properties [12–16]. The refractive index and extinction coefficient of each layer is broadly interpreted with spectroscopic ellipsometry measurements. Biswas et al. reported the ellipsometry studies of TiAlN/TiAlON/Si_3N_4 tandem absorber deposited on Cu substrate. They correlated the measured ellipsometry spectra with theoretical simulated spectra based on the optical constant determined for each layer [17]. Dan et al. reported the optical constants of W/WAlN/WAlON/Al_2O_3 multilayer coating with the presence of intermixed layers between WAlN and WAlON based on ellipsometry studies. The simulation using these optical constants indicates good correlation between the simulated and experimental measured reflectance spectra [18]. Similarly, Al-Rjoub et al. reported the optical constants of $WAlSiN_x$, $WAlSiO_yN_x$, and $SiAlO_y$ layers by varying the nitrogen and oxygen partial pressures. The multilayer stack was designed based on obtained optical constants using simulation software which exhibited a high solar absorptance of 0.96 and low thermal emissivity of 0.105 (calculated $\varepsilon_{400\ °C}$) [19]. Yet in another work, Escobar-Galindo et al. reported $Al_yTi_{1-y}(O_xN_{1-x})$ based solar absorber multilayer coatings, which are stable up to 650 °C for 12 h and the simulated results are in good accordance with the experimental reflectance data [20]. Similarly, Wang et al. reported an aperiodic metal-dielectric multilayer based on AlCrN and AlCrON high-temperature stable coating (500 °C for 1000 h) with high solar selectivity (α/ε) of 0.94/0.11 [21]. In this regard, a wide range of solar selective coatings developed by several groups are: Cu/TiAlCrN/TiAlN/AlSiN [22], $MoSi_2$–SiO_2 [23], Al-CrSiN/AlCrSiON/AlCrO [24], Al/NbMoN/NbMoON/SiO2 [25], TiN/nano-multilayered

AlCrSiO/amorphous AlCrSiO [26], and W/AlSiTiN$_x$/SiAlTiO$_y$N$_x$/SiAlO$_x$ [27].The optical design of the coating based on optical constants, layer thickness and composition plays a major role in developing a spectrally selective absorber coating with high solar absorptance and low thermal emissivity.

In our previous work, we have demonstrated the in-depth effect of process parameters affecting the optical properties of individual layers as well as the multilayer stack of W/WAlSiN/SiON/SiO_2 [28]. In this manuscript, we had presented the effect of reactive gas flow rates, sputtering power, deposition time and thicknesses of individual layers to achieve high spectral selectivity. A very low reflectance ($R < 5\%$) was observed in the wavelength range of 450–1500 nm and very high reflectance ($R > 95\%$) in the IR region, resulting in high solar absorptance ($\alpha = 0.955$) and low thermal emissivity ($\varepsilon = 0.10$ @ 82°C). In another study, the thermal stability of the optimized multilayer stack has been investigated in vacuum (700 °C for 200 h) and air (400 °C for 500 h and 500 °C for 100 h) for longer duration under cyclic heating conditions. Further, the high-temperature emissivity measurements have been carried out in the temperature range of 80 °C to 460 °C and the optimized sample exhibited a thermal emissivity of 0.15 @ 460 °C [29]. The proposed multilayer stack is a potential candidate as absorber coating on receiver tubes of parabolic through collector.

One of the biggest problems for central receiver tubes used in CSP is to improve the operating temperature of solar absorber coatings to enhance the overall photo-thermal conversion efficiency. In the literature, very few coatings are reported with good thermal stability (air) in the temperature range of 400–500 °C operating for long duration [20,23,27–30]. Because of high optical absorption along with low thermal emissivity as well as enhanced thermal stability of the tandem absorber of W/WAlSiN/SiON/SiO_2 owing to its unique nano-multilayer design, it is important to investigate the optical constants (n and k) of individual layers as well as tandem absorber. Therefore, in this manuscript, we report the phase-modulated spectroscopic ellipsometry measurements of individual layers of WAlSiN, SiON and SiO_2 and multilayer stack of W/WAlSiN/SiON/SiO_2 for their optical constants. The optical properties of thick individual layers were measured using UV-Vis spectroscopy and the effect of surface roughness on thermal emissivity was verified. The obtained refractive indices and extinction coefficients of individual layers and multilayers stack are curve fitted using suitable dispersion medium theories. The wide-angle absorptance was investigated by varying the incident angle in the UV-Vis-NIR region from 8° to 68°. The simulated reflectance spectra using obtained optical constants of the individual layers and multilayer stack are in good agreement with the experimentally measured reflectance spectra with minimal deviation.

2. Experimental Details

Spectrally selective coatings of W/WAlSiN/SiON/SiO_2 were deposited on stainless steel (SS) and silicon (Si) substrates by a Reactive Unbalanced Direct Current (DC) Magnetron Sputtering System with high purity (>99.9%) targets of W, Al, and Si. The substrate and targets were maintained at a constant distance of 10 cm throughout the deposition. Pulsed DC power supplies were used to deposit W, WAlSiN, SiON, and SiO_2 layers. The reactive sputtering of W, Al, Si targets in suitable Ar, N_2, and O_2 environments for depositing the multilayer stack and individual layers. All the coatings were deposited at a substrate temperature of 200 °C and in-situ Argon (Ar) plasma cleaning for 5 min at a voltage of −1000 V. The optimized process parameters (such as: Power density, reactive gas flow rates, bias voltage, and deposition time) were used to deposit each layers as discussed in our previous paper [28].

The refractive index (n) and the extinction coefficient (k) of the multilayer stack were measured by the phase-modulated spectroscopic ellipsometry (Model UVISEL 460, ISA Jobin-Yvon-Spex, Palaiseau, France) in the wavelength range of 300–900 nm. The obtained data were further analyzed by fitting with the appropriate dispersion models. Solar absorptance and emissivity of the as-deposited samples were measured using Solar Spectrum Reflectometer (Model SSR) and Emissometer (Model AE) (Devices & Services Company,

Texas, TX, USA) of Devices and Services. The emissivity was measured at a temperature of 82 °C. The reflectance spectra of the as-deposited samples were measured in the wavelength range of 0.25–2.5 μm using UV-VIS-NIR spectrophotometer (PerkinElmer: Lambda 950, (PerkinElmer, Massachusetts, MA, USA). Fourier transform infrared spectroscopy (PerkinElmer, Massachusetts, MA, USA) was used for measuring the reflectance from NIR to MIR (2–25 μm). Cross-sectional studies of the individual layers were carried out using field-emission scanning electron microscopy (FESEM, Carl Zeiss, SUPRA 40VP, Oberkochen, Germany). The cross-sectional FESEM images of individual layers were measured using secondary electron detection mode (SE) and at an acceleration voltage of 10 kV. The thicknesses of the individual layers were measured using a 3D profilometer, Nano Map500LS (AEP Technologies, California, CA, USA). A commercial simulation tool (SCOUT Version 2.99) was used for simulating the reflectance spectra of each layers of the tandem absorber using obtained optical constants and compared with the measured reflectance spectra [31].

3. Results and Discussion

3.1. Optical Properties of Individual Layers

The optical properties of individual layers in the multilayer stack enabled us to understand the spectral selectivity of the W/WAlSiN/SiON/SiO_2 solar absorber coatings. To obtain the individual layer optical properties, we have deposited the layers of WAlSiN, SiON and SiO_2 for long durations on SS and Si substrates. Their optical properties were measured in the wavelength range from 250–2500 nm using UV-Vis-NIR spectrophotometer. The thickness and average roughness of the film were measured from the Si substrate. The individual layer thicknesses were: 940, 445, and 400 nm for WAlSiN, SiON and SiO_2 layers, respectively, as labelled in insets of Figure 1. The layer thickness was measured using cross-section FESEM images as shown in the insets of Figure 1 and verified from 3D profilometer data. It is to be noted that the optical properties of W metal such as reflectance and refractive indices are well reported in the literature, therefore, the detailed characterization of W interlayer has not been carried out to avoid duplications. Figure 1a shows the reflectance spectrum of WAlSiN layer, wherein it is observed that the overall reflectance of the film is less than 40% in the UV-Vis-NIR region, indicating its absorbing nature. The layer has good absorption in UV-Vis-NIR region and we observe the interference fringes in near-IR region due constructive and destructive interference in the film. Similarly, the reflectance spectra of SiON and SiO_2 anti-reflection layers are as shown in Figure 1b,c with cross-section FESEM images of the coatings. However, the high reflectance of the SiON and SiO_2 layers indicates their non-absorbing nature. The SiON and SiO_2 layers are transparent in visible and near infrared region and acts as an excellent anti-reflection layers [32]. Moreover, the fringes observed in the UV-Vis region of reflectance spectra are due to interferences owing to the fact that light gets partially reflected from the substrate as these layers are transparent in this wavelength range.

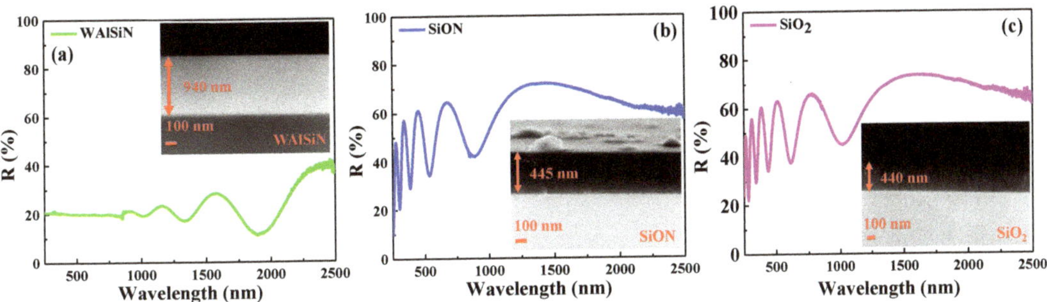

Figure 1. Reflectance spectra of individual layers deposited on stainless steel (SS) substrate (**a**) WAlSiN, (**b**) SiON, and (**c**) SiO_2. The cross-sectional FESEM images of the individual layers are shown in the insets.

The thermal emissivity depends on the material property and surface roughness of the coating. The coatings high surface roughness would considerably increase the thermal emissivity. The statement mentioned above has been proved theoretically and experimentally by various research groups [33]. The theoretical expression which correlates the dependence of thermal emissivity, surface roughness and reflectance of the coating is as in Equation (1):

$$R_r = R_p \exp\left\{-\left(\frac{4\pi\sigma}{\lambda}\right)^2\right\} \quad (1)$$

where R_r, R_p, σ and λ are reflectance of rough surface, reflectance of polished surfaces, root mean square roughness and wavelength, respectively. According to Kirchhoff's law, the emissivity and reflectance can be related using the following equation: $\varepsilon_\lambda = 1 - R_\lambda$ [34,35]. From the above two equations it is evident that with increasing surface roughness, the emissivity of the film increases [7]. However, a fine nanostructure formed on the coating surface will enhance the light absorption by multiple reflections due to light trapping [36]. Cao et al. numerically investigated the dependence of surface roughness of the film on the reflectance and thermal emissivity [37]. Similarly, Wen et al. reported the modelling of surface roughness of aluminum alloy by different theoretical models and results indicate a clear correlation between the surface roughness and thermal emissivity [38]. Recently, metal–liquid–crystal–metal (MLCM) based metasurface of Au/LC/Au are reported for their enhanced thermal camouflage by structuring the surface emissivity and optimizing the surface microstructure [39,40]. In this regard, we carried out AFM studies to evaluate the effect on optical properties based on surface roughness of individual layers and multilayer stack, as shown in Figure 2. The main absorber layer (WAlSiN, layer thickness ~940 nm) exhibited the average roughness (R_a) of 1.87 nm, as shown in Figure 2a and the layer exhibits a selectivity (α/ε) of 0.80/0.63. However, SiON and SiO_2 depicted low average roughness values of 1.31 and 1.16 nm, as shown in Figure 2b,c. The anti-reflection layers exhibit almost similar selectivity of 0.440/0.20 and 0.428/0.20 for SiON (layer thickness ~445 nm) and SiO_2 (layer thickness ~400 nm), respectively. The multilayer stack of W/WAlSiN/SiON/SiO_2 deposited on SS substrates depicts a low surface roughness of 0.61 nm and exhibits good spectral selectivity (α/ε) of 0.955/0.10. The results indicate that the surface roughness of the thin films influences the thermal emissivity of the coatings.

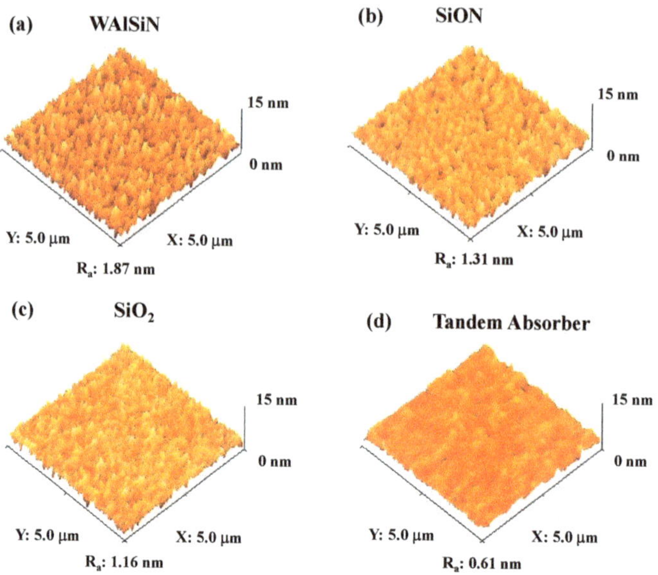

Figure 2. 3-dimensional AFM images of (**a**) WAlSiN, (**b**) SiON, (**c**) SiO_2, and (**d**) multilayer stack deposited on SS substrates.

3.2. Spectroscopy Ellipsometry Measurements

3.2.1. Optical Constants of Individual Layers

The solar selective properties of each layer were examined for a better understanding of optical properties. The individual layers of WAlSiN, SiON, and SiO_2 are deposited for measuring the optical constants using phase-modulated spectroscopic ellipsometry. In spectroscopic ellipsometry, it measures the change in the monochromatic light's polarization state reflected from the sample surface. The variations are represented as a function of ψ and Δ of the individual layers, which represents the amplitude ratio of s and p polarized light and their phase change, respectively. ψ and Δ can be represented in a complex reflection ratio (ρ), which is defined as the ratio of Fresnel reflection coefficient for s and p polarized light [17,41].

$$\rho = \frac{r_p}{r_s} = tan\psi \exp(i\Delta) \tag{2}$$

where, r_p and r_s are the reflection coefficient of p and s component of the electric field, respectively [17,18,41].

The obtained ψ and Δ data from ellipsometry are curve fitted by assuming a physical model appropriate for each layer. Theoretical simulated spectra consider a suitable optical dispersion medium for the layers. However, ψ and Δ data from ellipsometry measurements are derived from experiments and are curve fitted with an optical dispersion model for optical constant of individual layers. Some of the assumptions, considered during the fitting of experimentally obtained data with the theoretically generated data are [38]:

- The individual layer deposited are considered as homogenous model of thin-film and is simulated with a theoretical dispersion model to generate optical constants.
- The optical constants of the bare substrate are measured to attain more realistic results, and these results were compiled for curve fittings.

The optical constants are determined by considering a physical model that matches the sample and generates a generally suitable optical dispersion oscillator. In this regard, Cauchy's absorbent dispersion model was considered for WAlSiN main absorber layer, as this layer act as absorbing layer in the multilayer stack [17,18,42]. The relations for Cauchy's absorbent dispersion medium are as shown in Equations (3) and (4):

$$n(\lambda) = A + \frac{B}{\lambda^2} + \frac{C}{\lambda^4} \tag{3}$$

$$k(\lambda) = D + \frac{E}{\lambda^2} + \frac{F}{\lambda^4} \tag{4}$$

where, λ is the wavelength and A, B, C, D, E, and F are fit parameters. The above equation mentions that A, B, and C corresponds to the long-wavelength asymptotic refractive index value, slope and amplitude, respectively of refractive index curve, a similar notation for D, E, and F [17,18]. The top SiON and SiO_2 layers were curve fitted with the Tauc–Lorentz (TL) oscillator model. TL model is effective for nanocrystalline and amorphous thin films [18,43]. The complex dielectric functions can be expressed in a simple oscillator model and the expression for ε_2 is as shown in the Equation (5):

$$\varepsilon_2 = \frac{AE_o\Gamma\,(E - E_g)^2}{\left[(E - E_g)^2 + \Gamma^2 E^2\right]^2}\,\frac{1}{E};\ E > E_g \tag{5}$$

$$\varepsilon_2 = 0\,;\ E < E_g$$

where, E_o, E_g, Γ and A are the peak transition energy, optical band gap energy, broadening parameter, and optical transition matrix elements, respectively [20,44]. From Kramers–Krönig transformation (KKT) the real part (ε_1) of the dielectric function is expressed as depicted in Equation (6):

$$\varepsilon_1 = \varepsilon_{\alpha,UV} + \frac{2}{\pi}P\int_{E_b}^{\alpha}\frac{\xi\varepsilon_2(\xi)}{\xi^2 - E^2}d\xi \tag{6}$$

where, $\varepsilon_{\alpha,UV}$, ξ and P represents the high frequency dielectric constant, linear dielectric susceptibility and principal values of the integrals, respectively [45].

Using the above dispersion models ψ and Δ data of WAlSiN, SiON and SiO_2 samples were curve fitted using a minimization process. This minimization process has a maximum of 100 iterations and the basis of convergence is 0.000001 (χ^2 minimization—which defines the good fitting of curves). The model parameters are varied by a regression analysis until the calculated and experimental data are as close as possible. The following average square error function is minimized by weighing it to the approximate experimental errors.

$$\chi = \left\{\frac{1}{2N - M}\sum_{i=1}^{N}\left[\left(\frac{\psi_i^{mod} - \psi_i^{expt}}{\sigma_{\psi}^{expt},\ i}\right)^2 + \left(\frac{\Delta_i^{mod} - \Delta_i^{expt}}{\sigma_{\Delta}^{expt},\ i}\right)^2\right]\right\}^{\frac{1}{2}} \tag{7}$$

where, N, M, and σ are the number of measured ψ and Δ pairs, number of variable parameters and standard deviation, respectively. The superscripts "mod" and "expt" mean the theoretical calculations and experimental data [18,46]. The curve fitted ψ and Δ of individual layers are shown by lines, which are in good accordance with the measured data, represented by symbols in Figure 3. However, the oscillations observed in SiON and SiO_2 in the wavelength range of 300–900 nm, due to interference as shown in Figure 3b,c. In contrast, the absence of such oscillations for the WAlSiN layer, as depicted in Figure 3a, indicates a strongly absorbing property [19]. The optical behavior of such absorbing thin films can be observed from the reflectance spectra shown in Figure 1. It is to be noted that the WAlSiN layer contains fine nano-multilayers of W_2N and AlSiN, a total of 36 layers (18 layers each) as reported previously [29]. However, during the ellipsometry fitting we have considered the nano-multilayer structure of W_2N (t ~1.5 nm, polycrystalline phase) and AlSiN (t ~3 nm, amorphous phase) as a single composite layer of WAlSiN and effective optical constants of WAlSiN layer are obtained. This is assumed as the individual layer thickness (~3 nm) is very small compared to the measurement wavelength (300–900 nm) of the ellipsometry. The best fitted optical constant o are plotted as a function of wavelength, which are shown in Figure 4a. In the case of, WAlSiN layer the refractive index increases with wavelength, thereby making this layer ideal a material for selective absorption of solar radiation, as depicted in Figure 4a. The refractive indices of SiON and SiO_2 decrease with increasing wavelength, but the change is minimal throughout the wavelength range, as shown in Figure 4b,c. The decrease of the extinction coefficient (k) in WAlSiN film indicates the presence of interband transitions and metallic nature of the film. The "k" values of SiON and SiO_2 are zero as expected for dielectric materials. The excellent optical behavior of SiON and SiO_2 makes them a potential candidate for anti-reflection layers. The optical constants (n and k) of the individual layers were calculated using different dispersion models, discussed above.

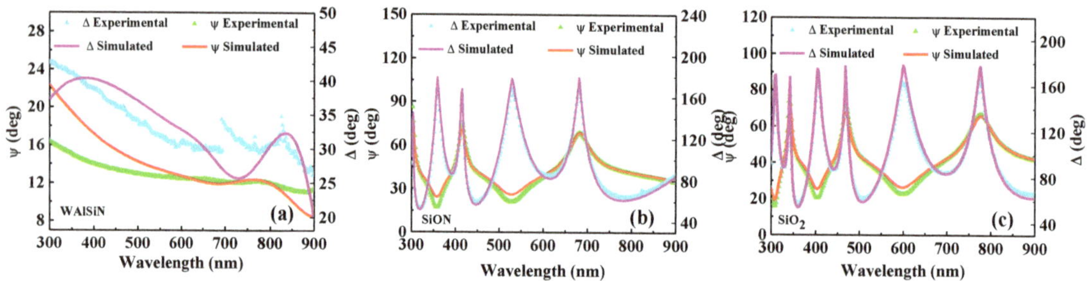

Figure 3. Ellipsometry spectra of individual layers, (**a**) WAlSiN, (**b**) SiON, and (**c**) SiO_2 deposited on SS substrate.

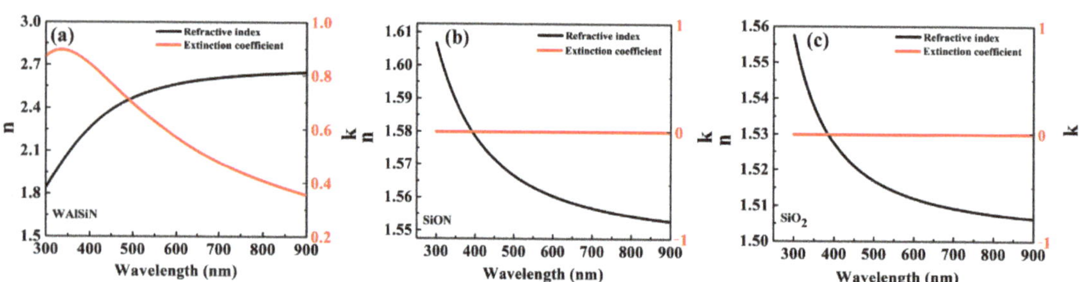

Figure 4. Refractive indices and extinction coefficients of individual layers (**a**) WAlSiN, (**b**) SiON, and (**c**) SiO_2 deposited on SS substrate.

The absorption coefficient of a material determines the depth of penetration of solar radiation into the material. It is well known that the extinction coefficient of a material is directly proportional to the optical absorption coefficient, which is as shown in the below equation:

$$\alpha = \frac{4\pi k}{\lambda} \tag{8}$$

where, α, k, and λ are absorption coefficient, extinction coefficient and wavelength [41]. The absorption coefficients calculated for individual layers using the obtained extinction coefficient are shown in Figure 5a. The higher absorption coefficient of the WAlSiN layer indicates it absorbs the incident solar radiation efficiently compared to that of SiON and SiO_2. Similarly, penetration depth indicates the extent to which the incident radiation penetrates inside the film thickness and the same for WAlSiN layer is shown in Figure 5b. For the WAlSiN layer with increasing wavelength, the penetration depth increases, indicating good absorption property of the layer. In contrast, SiON and SiO_2 films are transparent in the wavelength range of 300–900 nm.

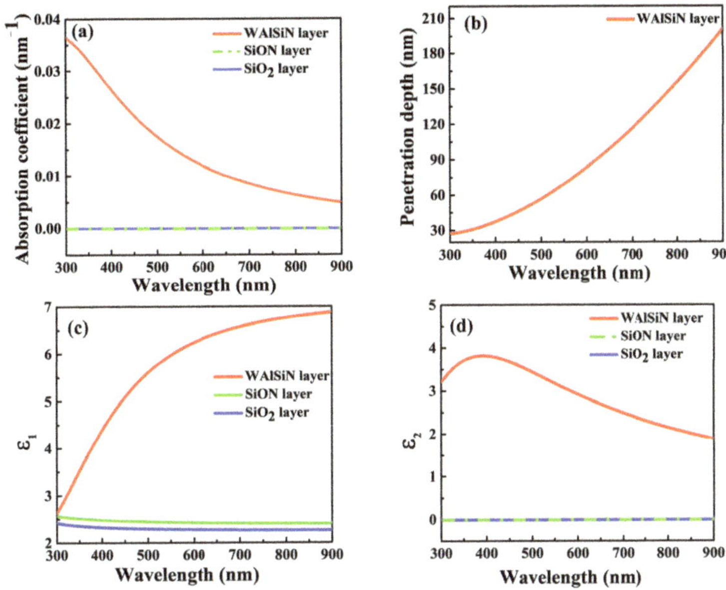

Figure 5. (**a**) Absorption coefficient, (**b**) penetration depth, (**c**) real (ε_1), and (**d**) imaginary (ε_2) part of the dielectric constants as a function of wavelength for single layer coatings deposited on SS substrate.

The optical properties of thin films are influenced by various material properties such as dielectric constant (ε), dielectric susceptibility (χ), and conductivity (σ), which are treated as scalars for isotropic materials. Moreover, semiconductors and dielectric films are considered nonmagnetic and do not possess excess electrons other than the electrons bound in atoms. The relation between the optical constants (n and k) for dielectric and semiconductor materials can be calculated from the following relations:

$$N = n + ik \text{ and } \varepsilon = \varepsilon_1 + i\,\varepsilon_2 \tag{9}$$

where, N and ε are complex refractive index and complex dielectric constant, respectively. Similarly, the real and imaginary part of dielectric constant can be calculated using $\varepsilon_1 = n^2 - k^2$ and $\varepsilon_2 = 2nk$, respectively [41]. The real (ε_1) and imaginary (ε_2) part of dielectric constants of individual layers were plotted using the obtained optical constants (n and k), which are shown in Figure 5c,d. The imaginary part (ε_2) of dielectric constant is related to the conductivity of the material and higher (ε_2) indicates good conductivity due to metallic nature of the film. The ε_1 is related to the polarization and ε_2 is related to dissipation, which accounts for wide-angle selectivity and better absorption in the film, respectively. The ε_2 for WAlSiN layers increases with wavelength, depicting excellent absorption property of the film. However, SiON and SiO_2 demonstrate no absorption and are highly transparent films in the wavelength range [47,48].

3.2.2. Optical Constants of Multilayer Stack

The optical properties of the optimized multilayer stack were designed based on gradation in optical constants of each layers calculated using the dispersion theories. The metal W is well explored and reported in the literature as a material with high refractive index of 3.83 and good IR reflector [49]. Moreover, the sole purpose of the W interlayer is to reduce the overall thermal emissivity of the coating and acts as a diffusion barrier at high temperature. The experimentally measured optical constants of individual layers were fitted for $W/WAlSiN/SiON/SiO_2$ with multilayer model. By considering this we account the overall optical behavior of the layers in the stack as well as the effect of each individual layers in achieving high solar selectivity (α/ε). The curve fitted ψ and Δ of the multilayer stack are shown in solid lines and which are in good accordance with the measured data, as shown in Figure 6a. The best curve fitting of the multilayer stack shown in Figure 6a emphasizes that the assumption of considering the nano-multilayers of W_2N and AlSiN layers present in WAlSiN main absorber layer, which is considered as a single composite layer, has insignificant impact on the calculated optical constants. The experimental fitting of the multilayer stack depicts the presence of intermixed layer in between WAlSiN and SiON layer. The optical constants of intermixed layer were evaluated using Bruggeman effective-medium approximation (EMA) [18,50,51]. The EMA model is sensitive to the surface roughness of the layers, but the intermixed layer thickness is ~5 nm and the surface roughness can be neglected. The intermixed layer formed at the interface of WAlSiN and SiON layer has to be better understood for its influence over the optical properties of the multilayer stack. The multilayer stacks optical constants with gradation in the refractive index and extinction coefficient are drawn as a function of wavelength and is as shown in Figure 6b,c. The refractive indices of W, WAlSiN, intermixed layer, SiON and SiO_2 in the multilayer stack measured at 550 nm are 3.83, 2.52, 2.28, 1.56, and 1.51, respectively, as shown in the inset of Figure 6b. The lower refractive index of SiON and SiO_2 layers shows the dielectric nature of the films. The trend of refractive index in the multilayer stack depicts an increase from top anti-reflection layers to the substrate. At each interface of the layers, the incident solar radiation will change the phase by 180°, leading to maximum absorption of light [52]. This graded refractive index concept is well reported and is an efficient way in trapping light and in achieving enhanced solar absorption of the multilayer coatings as explained in the below section [53–55].

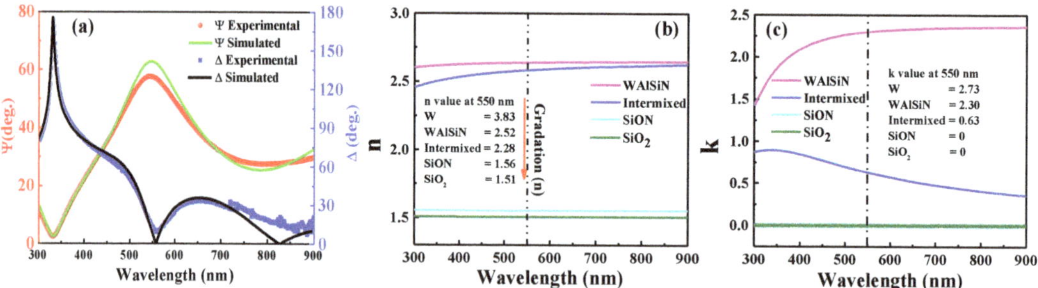

Figure 6. (**a**) Ellipsometry spectra of solar absorber coating, (**b**) refractive index and (**c**) extinction coefficient of $WAlSiN/SiON/SiO_2$ multilayer stack deposited on SS substrate. The refractive index and extinction coefficient of W are depicted in (**b**) and (**c**).

3.3. Optical Properties of Multilayer Stack

To understand the behavior of high solar absorptance of the multilayer stack, we have deposited successive layers step by step (SS, SS/W, SS/W/WAlSiN, SS/W/WAlSiN/SiON, and $SS/W/WAlSiN/SiON/SiO_2$) and measured the reflectance spectra of each sample. From the reflectance spectra it clear that by addition of one more layer on the top, the reflectance in the UV-Vis region reduces and near-zero reflectance is achieved after deposition of last layer, as shown in Figure 7. We have measured the reflectance of polished SS substrate as a reference of reflectance for characterization of layers deposited on it successively and the optical properties (α and ε) are tabulated in Table 1. The deposition of the WAlSiN layer with fine-nano multilayers over the W interlayer results in a significant drop of reflectance in the UV-Vis region due to interference, as seen in Figure 7. Further, adding of SiON and SiO_2 anti-reflection layers the reflectance to near zero in the wavelength range of 500–1300 nm is achieved due to gradation in refractive indices of the layers as shown in Figure 6 [56].

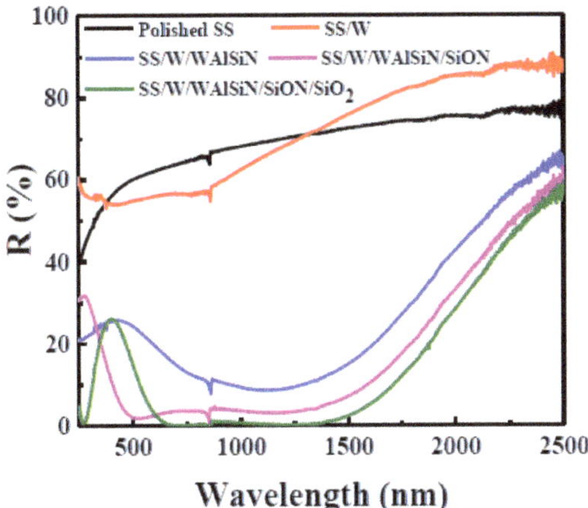

Figure 7. The reflectance spectra of successive layer-by-layer deposited on SS substrate.

Table 1. Solar absorptance and thermal emissivity of layer-by-layer deposited on SS substrate.

Sl. No	Description	Solar Absorptance (α)	Thermal Emissivity (ε)
1	SS	0.320	0.13
2	SS/W	0.40	0.08
3	SS/WAlSiN	0.850	0.12
4	SS/WAlSiN/SiON	0.950	0.11
5	SS/WAlSiN/SiON/SiO_2	0.955	0.10

The schematic representation of the multilayer tandem stack of W/WAlSiN/SiON/SiO_2 deposited on the SS substrate, as shown in Figure 8a. The schematic depicts an intermixed layer between WAlSiN and SiON layer, a detailed explanation of this is mentioned in the below section. The thicknesses of each individual layers in the multilayer stack are labelled in the schematic, as depicted in Figure 8a. The selective solar absorber coating is designed based on a graded refractive index with a double anti-reflection layer. The reflectance spectra of the multilayer stack are as shown in Figure 8b, exhibits near-zero reflectance in the wavelength range of 0.6–1.4 µm and high reflectance of above 90% in the infrared region [28]. The low reflectance and high absorptance in the multilayer stack is due to destructive interference and band-to-band transitions [3]. A graded design of a multilayer stack generates a step-by-step change in the refractive index, resulting in lower reflection due to interference effect. Additionally, the double anti-reflection layer (DLAR) of SiON/SiO_2 reduces the reflection losses at the surface and enhances the absorption by trapping the incident solar radiation [10,32,57]. Kim et al. reported the DLAR coatings of SiN_x/SiO_2 of different thicknesses and the refractive indices of each layer were theoretically calculated using Essential Macleod software. They also reported that DLAR coatings exhibited better solar efficiency when compared to single SiN_x anti-reflection layer [10]. Moreover, the absorption of light over a wide range of wavelength is achieved better in DLAR than a single ARC layer [58]. In summary, we have demonstrated that the gradient in the refractive index of the individual layers in the multilayer stack responsible for the enhanced absorption (high α).

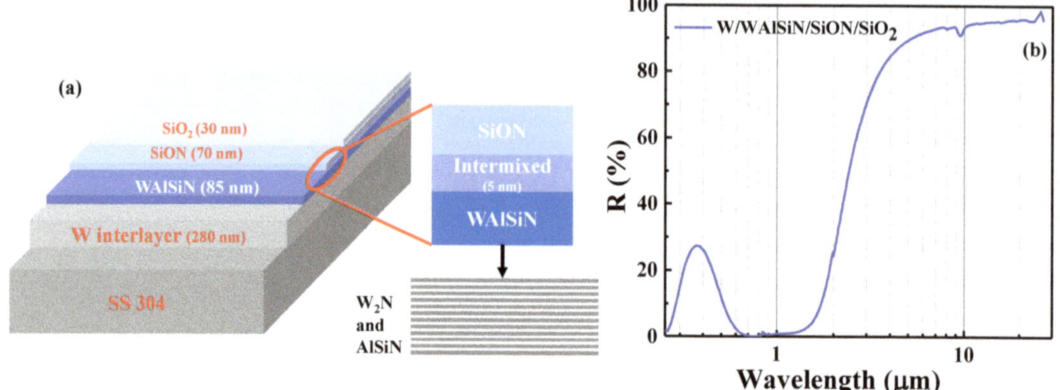

Figure 8. (**a**) Schematic of the multilayer stack deposited on SS substrate and (**b**) The reflectance spectrum of the optimized sample of the multilayer stack [28].

3.4. Angular Dependence of Solar Absorptance

The angle of incidence of solar radiation on the multilayer stack has direct influence over the reflectance and overall performance of the system. By varying the angle of incidence, the reflectance spectra are measured for the multilayer stack using UV-Vis-NIR spectrophotometer. The influence on optical properties of the multilayer stack with change in incident angle was studied in detail. The incident angle was varied from 8°–68° and the transverse electric (TE), i.e., *s* polarization and transverse magnetic (TM), i.e., p polarization reflectance spectra were recorded as a function of wavelength, as shown in Figure 9. The *p* polarization reflectance spectra (R_p) depict a decrease in reflectance up to an incident angle of 58° and further it increases at 68° incident angle, as shown in Figure 9a. The reflectance of the multilayer stack is less than 6% in the wavelength range of 500–1500 nm, which indicates good selectivity as well as wide angle solar absorptance. Similarly, the *s* polarization reflectance spectra (R_s) show an increase in reflectance of the film with increasing incident angle, as shown in Figure 9b. A slight shift in reflectance minima towards shorter wavelength is observed in reflectance spectra form 38°–68°. This is due the fact that at higher incident angles the effective thickness of the coating interacting with light is thinner compared to the actual thickness [59]. However, the average reflectance spectra of *s* and *p* polarization indicate a low reflectance (less than 10%) with varying incident angle, as shown in Figure 9c. These results demonstrate excellent wide-angle solar selectivity of W/WAlSiN/SiON/SiO_2 multilayer stack up to 58°.

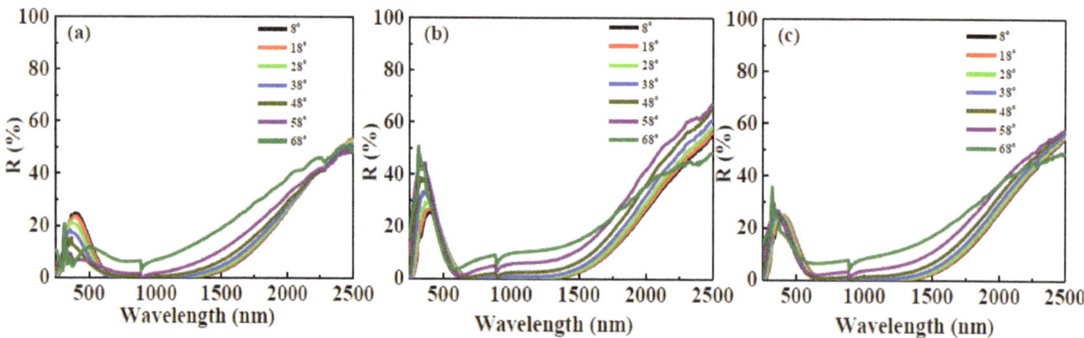

Figure 9. Angular absolute reflectance studies of the multilayer stack in UV-Vis-NIR region (**a**) TM polarization (R_p), (**b**) TE polarization (R_s), and (**c**) Average reflectance R = $(R_p + R_s)/2$.

3.5. Optical Simulation of Individual Layers and Multilayer Stack

A commercial simulation tool is used for determining the calculated reflectance spectra based on optical constants of the materials and film thicknesses [31]. The design of the solar absorber coating of W/WAlSiN/SiON/SiO_2 was carried out using simulation to minimize the number of experiments. The simulation was carried out to reduce the reflectance in the UV-ViS-NIR region by varying the refractive indices and thicknesses of the individual layers in the multilayer stack. Subsequently, the simulated reflectance data of the designed solar absorber coating was compared with the actual deposited W/WAlSiN/SiON/SiO_2 coating and similar approach was used for individual layers (WAlSiN, SiON and SiO_2) as well. The computational studies of WAlSiN, SiON, and SiO_2 layers were based on the optical constants obtained using spectroscopy ellipsometry and the reflectance spectra is generated in the wavelength range of 300–2500 nm. The refractive index (n) and extinction coefficient (k) of each individual layer are used for the simulation. Reference was inputted from experimentally measured reflectance spectra. The deviation was computed as a simple mean squared difference. Individual points on the simulated and imported reflectance spectra were compared and deviation was computed. The displayed value is the average of mean squared values over the defined range. The simulation fitting between the simulated and the measured spectra is evaluated using the deviation value. The simulation fitting is described as rejected, bad, acceptable, good and excellent based on the fit deviation value i.e., 0.1, 0.01, 0.001, 0.0001, and 0.00001, respectively [31]. The fit deviation values of all the simulated spectra are tabulated in Table 2 and a low deviation value indicates good and excellent fits of the simulated spectra. The simulated and experimental spectra of WAlSiN, SiON, and SiO_2 thick individual layers (Figure 10a–c and the deviations of the fit are tabulated in Table 2. The optical constants are acquired from ellipsometry studies of the individual layers as well as the multilayer stack, as discussed above. The data fitting in simulation software of the individual layers showed a deviation, which is attributed to available ellipsometry data of the individual layers in the limited wavelength range (300–900 nm). The deviation between the two spectra exists for various reasons. The most important one being the range over which ellipsometry data was collected. Ellipsometry data is collected up to 900 nm, while reflectance data is plotted up to 2500 nm. The simulation tool assumes a constant value (n, k at 900 nm) of the ellipsometry data in 900–2500 (missing data range). This can be justified as the fitting for the particular layer thickness well matched with the experimentally measured reflectance spectra in the range of 300–900 nm. These measurement errors influence a minor deviation in the reflectance spectra generated through SCOUT simulation software to that of the experimentally measured reflectance spectra. The fit deviations are tabulated in Table 2. However, the effective optical constant of WAlSiN layer used for simulation indicates an excellent fit, which implies that the calculated optical constants of WAlSiN layer are accurate.

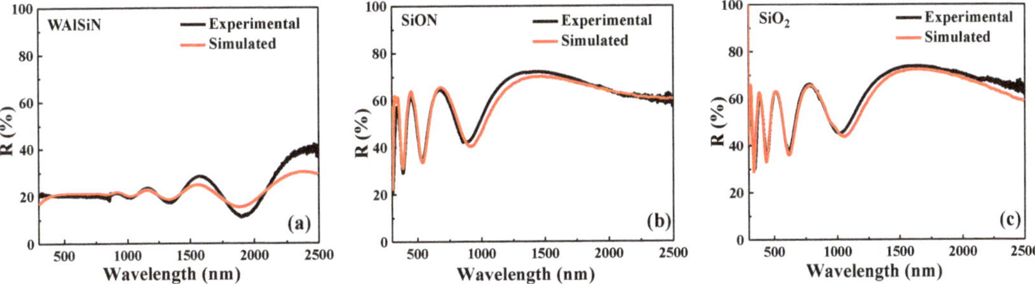

Figure 10. Experimental reflectance spectra of the individual layers in the tandem stack fitted with the simulated spectra obtained from SCOUT simulation software (**a**) WAlSiN, (**b**) SiON, and (**c**) SiO_2 layer.

Table 2. Thicknesses of the individual layers and tandem absorber used to generate the simulated reflectance spectra, fit deviation with experimentally measured spectra.

Sl. No	Description	Layer Thickness (nm)		Deviation
		Experimental	Simulated	
1	WAlSiN	940	872	0.00128
2	SiON	445	407	0.00371
3	SiO_2	440	430	0.00377
4	W/WAlSiN/SiON/SiO_2	280/85/70/30	250/78/68/37	0.0002270
5	W/WAlSiN/Intermixed layer/SiON/SiO_2	280/85/70/30	250/80/5/70/40	0.0002042

The effective optical constants of WAlSiN layer and optical constants of SiON and SiO_2 anti-reflection layers calculated are considered for simulation studies of the multilayer stack. In the multilayer stack, the simulation indicates an excellent fit with a deviation in fitting of 0.0002042 between the experimental and simulated spectra. Moreover, the thickness of individual layers in the multilayer stack after fitting shows a slight variation from the measured thickness of the film (as tabulated in Table 2), as this variation of thickness is due to the measured optical constants. However, at the interface of WAlSiN and SiON layer there exists a thin intermixed layer of ~5 nm in the multilayer stack, which is known from the spectroscopic ellipsometry measurement. The influences of the intermixed layer on reflectance spectra are simulated with respect to the measured reflectance spectra and the same is shown in Figure 11. The schematics of multilayer stack with simulated thicknesses are incorporated as in-sets in Figure 11a,b. The simulated spectrum without the presence of intermixed layer depicts a small hump in the wavelength range of 600–1200 nm and two reflectance minima at 712 and 1305 nm, as indicated in Figure 11a. The simulated thicknesses for the individual layers are labelled in the schematic (inset) of Figure 11a and are tabulated in Table 2. Figure 11b shows the simulated and the experimental reflectance spectra of multilayer stack in which the presence of the intermediate layer between WAlSiN main absorber layer and SiON layer exhibits an excellent fit. After introducing the intermixed layer, the reflectance minimum shifts from 712 nm to 637 nm, thereby implies better absorptance property of the multilayer stack, as marked in Figure 11b. However, the reflectance minimum at 1305 nm does not shift but the overall reflectance in the wavelength range of 600–1300 nm is less than 1%. Furthermore, the gradual change in the refractive index of the multilayer stack due to intermixed layer ensures the near-zero reflectance, i.e., W/WAlSiN/intermixedlayer/SiON/SiO_2 (3.83/2.52/2.28/1.56/1.51 @ 550 nm) from the top anti-reflection layer to the bottom of the substrate. This comparison of simulated spectra with and without intermixed layer confirms that due the formation of the intermixed layer in between WAlSiN and SiON layers, the reflectance decreases in the wavelength range of 600–1200 nm. So, the intermixing at the interface favors in lowering the reflectance by a gradual change in the optical constants. Similar, intermixing is also observed at the interface of WAlN and WAlON in W/WAlN/WAlON/Al_2O_3 solar selective coating [18]. Zhao et al. reported the optimized three-layer solar absorber coating using CODE simulation tool, which exhibits a high solar absorptance of 0.97 and low thermal emissivity [60]. Their simulated results were in good agreement with the experimental data, which indicates the reliability of the approach through computational simulation.

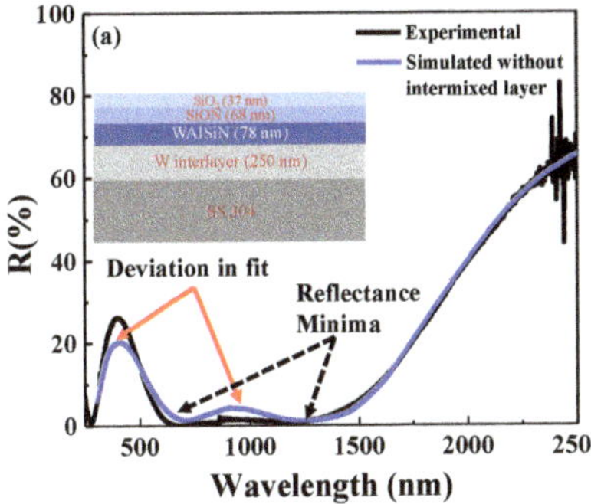

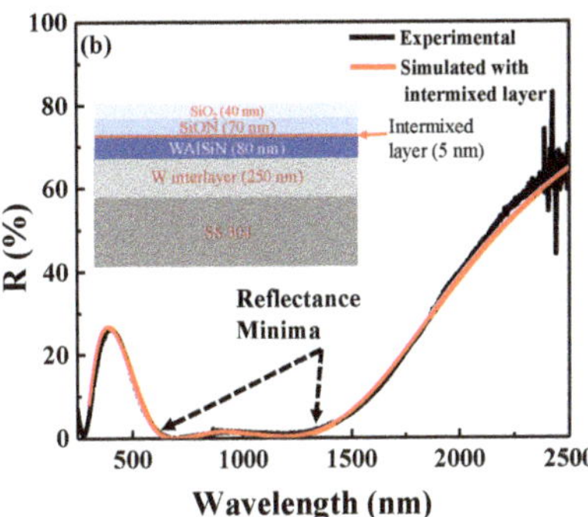

Figure 11. The multilayer reflectance spectra of the solar absorber fitted with simulated spectra generated from simulation: (**a**) Without the intermixed layer and (**b**) with the intermixed layer. The schematics of the multi-layer stack with simulated layer thicknesses are presented as insets.

4. Conclusions

We report the detailed evaluation and analysis of enhanced spectrally selective solar absorber coating of W/WAlSiN/SiON/SiO_2 by means of spectroscopic ellipsometry. The optical constants of individual layers and multilayer stack were obtained by curve fitting the ψ and Δ plots using appropriate dispersion models. The Cauchy absorbent dispersion model was used for main absorber layer (WAlSiN), Tauc-Lorentz model for anti-reflection layers (SiON and SiO_2) and Bruggeman effective medium approximation for the intermixed layer formed in between WAlSiN and SiON layers. The obtained optical constants depict a gradation in the designed multilayer stack from top anti-reflection layer to substrate due to which an enhanced solar absorption is achieved. The multilayer stack of W/WAlSiN/SiON/SiO_2 (280/85/70/30 nm) deposited on SS substrate exhibits a high solar absorptance of 0.955 and low thermal emissivity of 0.10. Furthermore, the wide-angle selectivity of the multilayer stack was measured by varying the incident angle (8°–68°) and the multilayer stack exhibits an excellent wide-angle selectivity up to 58°. The optical simulated spectra of individual layers and multilayer stack using software shows a good correlation between the measured and experimental spectra. The simulation results based on optical constants of multilayer stack measured using ellipsometry describe the influence of intermixed layer in achieving high solar absorptance.

Author Contributions: Conceptualization, H.C.B.; methodology, K.N.; software, K.N. and P.K.; validation, K.N., P.K., and A.B.; formal analysis, V.P.K. and G.S.; investigation, K.N.; resources, H.C.B.; data curation, K.N. and A.B.; writing—original draft preparation, K.N; writing—review and editing, H.C.B.; visualization, H.C.B. and K.N.; supervision, H.C.B.; project ad-ministration, H.C.B.; funding acquisition, H.C.B. All authors have read and agreed to the published version of the manuscript.

Funding: This research was funded by Department of Science and Technology (DST), grant number DST (U-1-144) and Centre Franco-Indien pour la Promotion de la Recherche Avancée (CEFIPRA), grant number (U-1-154).

Acknowledgments: The authors thank Siju John for FESEM measurement measurements. We thank the Department of Science and Technology, DST (U-1-144) and Centre Franco-Indien pour la Promotion de la Recherche Avancée, CEFIPRA (U-1-154) for providing funding support. K. Niranjan thanks CSIR for providing CSIR-SRF fellowship.

Conflicts of Interest: All authors declare no conflict of interest.

References

1. Weinstein, L.A.; Loomis, J.; Bhatia, B.S.; Bierman, D.M.; Wang, E.N.; Chen, G. Concentrating solar power. *Chem. Rev.* **2015**, *115*, 12797–12838. [CrossRef]
2. Kennedy, C.E. *Review of Mid-to High-Temperature Solar Selective Absorber Materials*; No. NREL/TP-520-31267; National Renewable Energy Lab.: Golden, CO, USA, July 2002.

3. Zhang, Q.-C. Recent progress in high-temperature solar selective coatings. *Sol. Energy Mater. Sol. Cells* **2000**, *62*, 63–74. [CrossRef]
4. Selvakumar, N.; Barshilia, H.C. Review of physical vapor deposited (PVD) spectrally selective coatings for mid-and high-temperature solar thermal applications. *Sol. Energy Mater. Sol. Cells* **2012**, *98*, 1–23. [CrossRef]
5. Xu, K.; Du, M.; Hao, L.; Mi, J.; Yu, Q.; Li, S. A review of high-temperature selective absorbing coatings for solar thermal applications. *J. Mater.* **2020**, *6*, 167–182. [CrossRef]
6. Cao, F.; Kraemer, D.; Sun, T.; Lan, Y.; Chen, G.; Ren, Z. Enhanced thermal stability of W-Ni-Al2O3Cermet-based spectrally selective solar absorbers with tungsten infrared reflectors. *Adv. Energy Mater.* **2015**, *5*, 1401042. [CrossRef]
7. Sibin, K.; John, S.; Barshilia, H.C. Control of thermal emittance of stainless steel using sputtered tungsten thin films for solar thermal power applications. *Sol. Energy Mater. Sol. Cells* **2015**, *133*, 1–7. [CrossRef]
8. Ibrahim, K.; A Taha, H.; Rahman, M.M.; Kabir, H.; Jiang, Z.-T. Solar selective performance of metal nitride/oxynitride based magnetron sputtered thin film coatings: A comprehensive review. *J. Opt.* **2018**, *20*, 033001. [CrossRef]
9. Kanda, H.; Uzum, A.; Harano, N.; Yoshinaga, S.; Ishikawa, Y.; Uraoka, Y.; Fukui, H.; Harada, T.; Ito, S. Al_2O_3/TiO_2 double layer anti-reflection coating film for crystalline silicon solar cells formed by spray pyrolysis. *Energy Sci. Eng.* **2016**, *4*, 269–276. [CrossRef]
10. Kim, J.; Park, J.; Hong, J.H.; Choi, S.J.; Kang, G.H.; Yu, G.J.; Kim, N.S.; Song, H.-E. Double antireflection coating layer with silicon nitride and silicon oxide for crystalline silicon solar cell. *J. Electroceram.* **2012**, *30*, 41–45. [CrossRef]
11. Rebouta, L.; Sousa, A.; Andritschky, M.; Cerqueira, F.; Tavares, C.; Santilli, P.; Pischow, K. Solar selective absorbing coatings based on AlSiN/AlSiON/$AlSiO_y$ layers. *Appl. Surf. Sci.* **2015**, *356*, 203–212. [CrossRef]
12. Barshilia, H.C.; Selvakumar, N.; Rajam, K.; Biswas, A. Spectrally selective NbAlN/NbAlON/Si3N4 tandem absorber for high-temperature solar applications. *Sol. Energy Mater. Sol. Cells* **2008**, *92*, 495–504. [CrossRef]
13. Bilokur, M.; Gentle, A.; Arnold, M.D.; Cortie, M.B.; Smith, G.B. Spectrally selective solar absorbers based on Ta:SiO_2 Cermets for next-generation concentrated solar-thermal applications. *Energy Technol.* **2020**, *8*, 2000125. [CrossRef]
14. Snyder, P.G.; Xiong, Y.; Woollam, J.A.; Al-Jumaily, G.A.; Gagliardi, F.J. Graded refractive index silicon oxynitride thin film characterized by spectroscopic ellipsometry. *J. Vac. Sci. Technol. A* **1992**, *10*, 1462–1466. [CrossRef]
15. Al-Rjoub, A.; Rebouta, L.; Costa, P.; Barradas, N.; Alves, E.; Ferreira, P.; Abderrafi, K.; Matilainen, A.; Pischow, K. A design of selective solar absorber for high temperature applications. *Sol. Energy* **2018**, *172*, 177–183. [CrossRef]
16. Barshilia, H.C.; Selvakumar, N.; Rajam, K.S.; Rao, D.V.S.; Muraleedharan, K.; Biswas, A. TiAlN/TiAlON/Si3N4 tandem absorber for high temperature solar selective applications. *Appl. Phys. Lett.* **2006**, *89*, 191909. [CrossRef]
17. Biswas, A.; Bhattacharyya, D.; Barshilia, H.; Selvakumar, N.; Rajam, K. Spectroscopic ellipsometric characterization of TiAlN/TiAlON/Si3N4 tandem absorber for solar selective applications. *Appl. Surf. Sci.* **2008**, *254*, 1694–1699. [CrossRef]
18. Dan, A.; Biswas, A.; Sarkar, P.; Kashyap, S.; Chattopadhyay, K.; Barshilia, H.C.; Basu, B. Enhancing spectrally selective response of W/WAlN/WAlON/Al2O3-based nanostructured multilayer absorber coating through graded optical constants. *Sol. Energy Mater. Sol. Cells* **2018**, *176*, 157–166. [CrossRef]
19. Al-Rjoub, A.; Rebouta, L.; Costa, P.; Vieira, L. Multi-layer solar selective absorber coatings based on W/WSiAlNx /WSiAlO-yNx/SiAlOx for high temperature applications. *Sol. Energy Mater. Sol. Cells* **2018**, *186*, 300–308. [CrossRef]
20. Escobar-Galindo, R.; Guillén, E.; Heras, I.; Rincón-Llorente, G.; Alcón-Camas, M.; Lungwitz, F.; Munnik, F.; Schumann, E.; Azkona, I.; Krause, M. Design of high-temperature solar-selective coatings based on aluminium titanium oxynitrides AlyTi1-y(OxN1-x). Part 2: Experimental validation and durability tests at high temperature. *Sol. Energy Mater. Sol. Cells* **2018**, *185*, 183–191. [CrossRef]
21. Wang, X.; Luo, T.; Li, Q.; Cheng, X.; Li, K. High performance aperiodic metal-dielectric multilayer stacks for solar energy thermal conversion. *Sol. Energy Mater. Sol. Cells* **2019**, *191*, 372–380. [CrossRef]
22. Valleti, K.; Krishna, D.M.; Joshi, S. Functional multi-layer nitride coatings for high temperature solar selective applications. *Sol. Energy Mater. Sol. Cells* **2014**, *121*, 14–21. [CrossRef]
23. Liu, Y.; Wu, Z.; Yin, L.; Zhang, Z.; Wu, X.; Wei, D.; Zhang, Q.; Cao, F. High-temperature air-stable solar absorbing coatings based on the cermet of MoSi2 embedded in SiO2. *Sol. Energy Mater. Sol. Cells* **2019**, *200*, 109946. [CrossRef]
24. Zou, C.; Xie, W.; Shao, L. Functional multi-layer solar spectral selective absorbing coatings of AlCrSiN/AlCrSiON/AlCrO for high temperature applications. *Sol. Energy Mater. Sol. Cells* **2016**, *153*, 9–17. [CrossRef]
25. Song, P.; Wu, Y.; Wang, L.; Sun, Y.; Ning, Y.; Zhang, Y.; Dai, B.; Tomasella, E.; Bousquet, A.; Wang, C. The investigation of thermal stability of Al/NbMoN/NbMoON/SiO2 solar selective absorbing coating. *Sol. Energy Mater. Sol. Cells* **2017**, *171*, 253–257. [CrossRef]
26. Yang, D.; Zhao, X.; Liu, Y.; Li, J.; Liu, H.; Hu, X.; Li, Z.; Zhang, J.; Guo, J.; Chen, Y.; et al. Enhanced thermal stability of solar selective absorber based on nano-multilayered AlCrSiO films. *Sol. Energy Mater. Sol. Cells* **2020**, *207*, 110331. [CrossRef]
27. Al-Rjoub, A.; Rebouta, L.; Cunha, N.; Fernandes, F.; Barradas, N.; Alves, E. W/AlSiTiNx/SiAlTiOyNx/SiAlOx multilayered solar thermal selective absorber coating. *Sol. Energy* **2020**, *207*, 192–198. [CrossRef]
28. Niranjan, K.; Kondaiah, P.; Srinivas, G.; Barshilia, H.C. Optimization of W/WAlSiN/SiON/SiO2 tandem absorber consisting of double layer anti-reflection coating with broadband absorption in the solar spectrum region. *Appl. Surf. Sci.* **2019**, *496*, 143651. [CrossRef]

29. Niranjan, K.; Soum-Glaude, A.; Carling-Plaza, A.; Bysakh, S.; John, S.; Barshilia, H.C. Extremely high temperature stable nanometric scale multilayer spectrally selective absorber coating: Emissivity measurements at elevated temperatures and a comprehensive study on ageing mechanism. *Sol. Energy Mater. Sol. Cells* **2021**, *221*, 110905. [CrossRef]
30. Manikandan, G.; Iniyan, S.; Goic, R. Enhancing the optical and thermal efficiency of a parabolic trough collector—A review. *Appl. Energy* **2019**, *235*, 1524–1540. [CrossRef]
31. *SCOUT Thin Film Analysis Software Handbook*; W Theiss Hard and Software: Aachen, Germany, 2017.
32. Meziani, S.; Moussi, A.; Mahiou, L.; Outemzabet, R. Effect of thermal annealing on double anti reflection coating SiNx/SiO2. In Proceedings of the 2015 3rd International Renewable and Sustainable Energy Conference (IRSEC), Marrakech, Morocco, 10–13 December 2015; pp. 1–5.
33. He, X.; Li, Y.; Wang, L.; Sun, Y.; Zhang, S. High emissivity coatings for high temperature application: Progress and prospect. *Thin Solid Films* **2009**, *517*, 5120–5129. [CrossRef]
34. Hu, R.; Song, J.; Liu, Y.; Xi, W.; Zhao, Y.; Yu, X.; Cheng, Q.; Tao, G.; Luo, X. Machine learning-optimized Tamm emitter for high-performance thermophotovoltaic system with detailed balance analysis. *Nano Energy* **2020**, *72*, 104687. [CrossRef]
35. Xi, W.; Liu, Y.; Song, J.; Hu, R.; Luo, X. High-throughput screening of a high-Q mid-infrared Tamm emitter by material informatics. *Opt. Lett.* **2021**, *46*, 888–891. [CrossRef] [PubMed]
36. Raut, H.K.; Ganesh, V.A.; Nair, A.S.; Ramakrishna, S. Anti-reflective coatings: A critical, in-depth review. *Energy Environ. Sci.* **2011**, *4*, 3779–3804. [CrossRef]
37. Cao, L.; Sendur, K. Surface roughness effects on the broadband reflection for refractory metals and polar dielectrics. *Materials* **2019**, *12*, 3090. [CrossRef]
38. Wen, C.-D.; Mudawar, I. Modeling the effects of surface roughness on the emissivity of aluminum alloys. *Int. J. Heat Mass Transf.* **2006**, *49*, 4279–4289. [CrossRef]
39. Liu, Y.; Song, J.; Zhao, W.; Ren, X.; Cheng, Q.; Luo, X.; Fang, N.X.; Hu, R. Dynamic thermal camouflage via a liquid-crystal-based radiative metasurface. *Nanophotonics* **2020**, *9*, 855–863. [CrossRef]
40. Song, J.; Huang, S.; Ma, Y.; Cheng, Q.; Hu, R.; Luo, X. Radiative metasurface for thermal camouflage, illusion and messaging. *Opt. Express* **2020**, *28*, 875–885. [CrossRef] [PubMed]
41. Fujiwara, H. *Spectroscopic Ellipsometry: Principles and Applications*; John Wiley & Sons, Ltd.: Chichester, England, 2007.
42. Knopp, K.J.; Mirin, R.P.; Bertness, K.A.; Silverman, K.L.; Christensen, D.H. Compound semiconductor oxide antireflection coatings. *J. Appl. Phys.* **2000**, *87*, 7169–7175. [CrossRef]
43. Hilfiker, J.N.; Singh, N.; Tiwald, T.; Convey, D.; Smith, S.M.; Baker, J.H.; Tompkins, H.G. Survey of methods to characterize thin absorbing films with spectroscopic ellipsometry. *Thin Solid Films* **2008**, *516*, 7979–7989. [CrossRef]
44. Jellison, G.E., Jr.; Modine, F.A. Parameterization of the optical functions of amorphous materials in the interband region. *Appl. Phys. Lett.* **1996**, *69*, 371–373. [CrossRef]
45. Jellison, J.G.; Modine, F.; Doshi, P.; Rohatgi, A. Spectroscopic ellipsometry characterization of thin-film silicon nitride. *Thin Solid Films* **1998**, *313-314*, 193–197. [CrossRef]
46. Jellison, J.G. Spectroscopic ellipsometry data analysis: Measured versus calculated quantities. *Thin Solid Films* **1998**, *313–314*, 33–39. [CrossRef]
47. Wemple, S.H. Refractive-index behavior of amorphous semiconductors and glasses. *Phys. Rev. B* **1973**, *7*, 3767–3777. [CrossRef]
48. Ma, H.-P.; Lu, H.-L.; Yang, J.-H.; Li, X.-X.; Wang, T.; Huang, W.; Yuan, G.-J.; Komarov, F.F.; Zhang, D.W. Measurements of microstructural, chemical, optical, and electrical properties of silicon-oxygen-nitrogen films prepared by plasma-enhanced atomic layer deposition. *Nanomaterials* **2018**, *8*, 1008. [CrossRef]
49. Ordal, M.A.; Long, L.L.; Bell, R.J.; Bell, S.E.; Alexander, R.W.; Ward, C.A. Optical properties of the metals Al, Co, Cu, Au, Fe, Pb, Ni, Pd, Pt, Ag, Ti, and W in the infrared and far infrared. *Appl. Opt.* **1983**, *22*, 1099–1119. [CrossRef]
50. Sancho-Parramon, J.; Janicki, V.; Zorc, H. On the dielectric function tuning of random metal-dielectric nanocomposites for metamaterial applications. *Opt. Express* **2010**, *18*, 26915–26928. [CrossRef] [PubMed]
51. Heras, I.; Krause, M.; Abrasonis, G.; Pardo, A.; Endrino, J.; Guillén, E.; Escobar-Galindo, R. Advanced characterization and optical simulation for the design of solar selective coatings based on carbon: Transition metal carbide nanocomposites. *Sol. Energy Mater. Sol. Cells* **2016**, *157*, 580–590. [CrossRef]
52. Keçebaş, M.A.; Şendur, K. Enhancing the spectral reflectance of refractory metals by multilayer optical thin-film coatings. *J. Opt. Soc. Am. B* **2018**, *35*, 1845–1853. [CrossRef]
53. Ji, D.; Song, H.; Zeng, X.; Hu, H.; Liu, K.; Zhang, N.; Gan, Q. Broadband absorption engineering of hyperbolic metafilm patterns. *Sci. Rep.* **2015**, *4*, 4498. [CrossRef] [PubMed]
54. Joly, M.; Antonetti, Y.; Python, M.; Lazo, M.G.; Gascou, T.; Hessler-Wyser, A.; Scartezzini, J.-L.; Schüler, A. Selective solar absorber coatings on receiver tubes for CSP—Energy-efficient production process by sol-gel dip-coating and subsequent induction heating. *Energy Procedia* **2014**, *57*, 487–496. [CrossRef]
55. Rebouta, L.; Capela, P.; Andritschky, M.; Matilainen, A.; Santilli, P.; Pischow, K.; Alves, E. Characterization of TiAlSiN/TiAlSiON/SiO2 optical stack designed by modelling calculations for solar selective applications. *Sol. Energy Mater. Sol. Cells* **2012**, *105*, 202–207. [CrossRef]
56. Soum-Glaude, A.; Di Giacomo, L.; Quoizola, S.; Laurent, T.; Flamant, G. Selective surfaces for solar thermal energy conversion in CSP: From multilayers to nanocomposites. *Nanotechnol. Energy Sustain.* **2017**, 231–248. [CrossRef]

57. Li, M.; Zeng, L.; Chen, Y.; Zhuang, L.; Wang, X.; Shen, H. Realization of colored multicrystalline silicon solar cells with SiO2/SiNx:H double layer antireflection coatings. *Int. J. Photoenergy* **2013**, *2013*, 352473. [CrossRef]
58. Zhang, J.; Chen, T.; Liu, Y.; Liu, Z.; Yang, H. Modeling of a selective solar absorber thin film structure based on double TiNxOy layers for concentrated solar power applications. *Sol. Energy* **2017**, *142*, 33–38. [CrossRef]
59. Zheng, L.; Zhou, F.; Zhou, Z.; Song, X.; Dong, G.; Wang, M.; Diao, X. Angular solar absorptance and thermal stability of Mo–SiO_2 double cermet solar selective absorber coating. *Sol. Energy* **2015**, *115*, 341–346. [CrossRef]
60. Zhao, S.; Wäckelgård, E. Optimization of solar absorbing three-layer coatings. *Sol. Energy Mater. Sol. Cells* **2006**, *90*, 243–261. [CrossRef]

Article

Multilayer Nanoimprinting to Create Hierarchical Stamp Masters for Nanoimprinting of Optical Micro- and Nanostructures

Amiya R. Moharana, Helene M. Außerhuber, Tina Mitteramskogler, Michael J. Haslinger and Michael M. Mühlberger *

PROFACTOR GmbH, Im Stadtgut A2, 4407 Steyr-Gleink, Austria; amiya.moharana@profactor.at (A.R.M.); helene.ausserhuber@profactor.at (H.M.A.); tina.mitteramskogler@profactor.at (T.M.); michael.haslinger@profactor.at (M.J.H.)

* Correspondence: michael.muehlberger@profactor.at

Received: 28 February 2020; Accepted: 17 March 2020; Published: 24 March 2020

Abstract: Nanoimprinting is a well-established replication technology for optical elements, with the capability to replicate highly complex micro- and nanostructures. One of the main challenges, however, is the generation of the master structures necessary for stamp fabrication. We used UV-based Nanoimprint Lithography to prepare hierarchical master structures. To realize structures with two different length scales, conventional nanoimprinting of larger structures and conformal reversal nanoimprinting to print smaller structures on top of the larger structures was performed. Liquid transfer imprint lithography proved to be well suited for this purpose. We used the sample prepared in such a way as a master for further nanoimprinting, where the hierarchical structures can then be imprinted in one single nanoimprinting step. As an example, we presented a diffusor structure with a diffraction-grating structure on top.

Keywords: Nanoimprint lithography; UV-NIL; reversal NIL; liquid transfer imprint lithography; hierarchical structures; optical micro- and nanostructures

1. Introduction

UV-based Nanoimprint Lithography [1,2] is a method to replicate nanostructures on a large area, and is commonly used for the fabrication of microlenses [3–5], wafer-level cameras [6,7] and diffractive optical elements [8–10]. The basic process flow is the following: a nanostructured stamp is pressed into a liquid, UV-curable material on a substrate. While the stamp is in contact with the polymer, the material is cured by UV-irradiation and then the stamp is removed, resulting in a nanostructured crosslinked polymer on the substrate. Typically, and for cost-of-ownership considerations, the stamp itself is a copy from a master structure, which can be prepared by many different types of fabrication methods, for example, electron beam lithography [11,12], ion beam lithography [13], x-ray lithography [14], diamond turning [10,15] and even by copying from natural structures [16]. For many applications, it is interesting to investigate nanostructures on top of larger microstructures: so-called 'hierarchical structures'. Such structures often appear in nature (e.g., lotus flower leaves [17] or gecko feet [18,19]), but can also be interesting for other applications like grating couplers on waveguides (e.g., [20]), energy applications [21], or sensors [22]). Nanoimprinting is ideally suited to replicate complex structures in a single process step. In most cases it is, however, non-trivial to fabricate the complex master structure, which is then used for stamp fabrication. We investigated the combination of two nanoimprint steps to create such masters with hierarchical structures.

2. Materials and Methods

As nanoimprint materials, we focused, in this work, on UV-curable hybrid polymers, so-called 'Ormocers'. They are commercially available and are designed for use in optical applications [23].

Usually, in nanoimprint lithography, it is the substrate which is coated with the UV-curable imprint polymer, e.g., by spin coating, droplet dispensing or inkjet printing. However, for certain applications, it can be advantageous to coat the stamp rather than the substrate, which is then called 'reversal NIL' [24]. This procedure is especially advantageous if NIL has to be performed on top of existing micro- or nanostructures while preserving the topography. It is usually not possible to apply a homogeneous coating directly on top of already existing micro- or nanostructures without filling up the gap between these structures or changing their geometry. However, reversal NIL is often challenging since the Nanoimprint stamps are designed to exhibit anti-sticking properties, which makes good wetting - and therefore homogeneous coating of the stamp – difficult, or leads to spontaneous dewetting. This puts limitations on the type of materials that can be used as well as on the type of stamps.

A variant of reversal NIL, so-called liquid-transfer imprint lithography (LTIL) [25], is a possible solution for this problem. The basic principle of LTIL is that, initially, a flat substrate is spin coated with the UV-curable resin. This so-called donor wafer is a material that can be spin-coated nicely. It is used to prepare a thin and well defined nanoimprint material layer on a substrate. Next, a flexible stamp is brought into contact with the resin layer on the donor wafer, and this stamp is then peeled off from the donor wafer in a well-controlled way. A thin, homogeneous layer of liquid UV-curable resin stays on the stamp, which is then transferred to the target substrate, where the UV-NIL process is finished by curing the resin by UV radiation and the subsequent demolding step.

This LTIL process can be used to nanoimprint a structure on an already prestructured substrate to combine different structures and functionalities [26–28]. In our work, we combined the optical effects of a line and space diffraction grating with that of a diffusor in order to create a sample with such a hierarchical pattern, whichwe then used as a new master for further stamp fabrication and imprinting, improving the facilitation of the overall process in the end. The diffusor and grating pattern was chosen to demonstrate the technology and has no specific application in the context of this work.

In the following paragraphs we describe the way in which we prepared our prestructured substrates, how we perform the LTIL nanoimprinting on top of the prestructured substrate, and how we used the result of this process as a master for working stamp fabrication and single-step imprinting of hierarchical structures.

As the material for our prestructured substrates, we chose OrmoComp [29] on a standard PVC foil. The 100-μm-thick PVC foils were used as received, without any further treatment. OrmoComp has good adhesion to PVC and excellent optical and nanoimprint properties. We used a standard nanoimprint process to pattern the OrmoComp and create an OrmoComp diffusor substrate. The stamps for this nanoimprint process were prepared by casting Polydimethylsiloxane (PDMS, Sylgard 184, 1:10 mixing ratio, 40 °C temperature, 23 h curing time) on a commercial diffusor foil. The imprint material was droplet dispensed onto the PVC substrate. UV curing was accomplished using an in-house built high-power UV-LED light source with a wavelength of 365 nm. For the imprinting, no pressure was applied. A photograph and an AFM image are shown in Figure 1. The typical peak to valley height is around 2.5 μm.

On top of the OrmoComp diffusor substrate, we performed the LTIL nanoimprint process. The stamps for this process were also made from PDMS (Sylgard 184, same procedure as for the diffusor PDMS stamp) and copied from a line and space quartz master (fabricated by electron beam lithography and reactive ion etching and treated with our BGL-GZ-83 anti adhesion layer [30,31]). Figure 2 shows a photograph and an AFM image of the master that was used. It contains, in total, 16 4×4 mm^2 large fields with four different periodicities. The line and space dimensions are 600 nm/1800 nm, 400 nm/1200 nm, 300 nm/900 nm and 200 nm/600 nm.

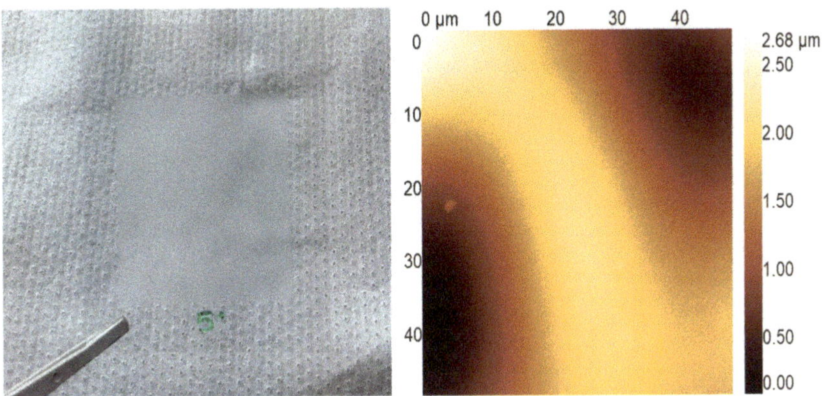

Figure 1. Left: photograph of the diffusor foil master. Right: AFM image of this master. The scan area is $50 \times 50\ \mu m^2$.

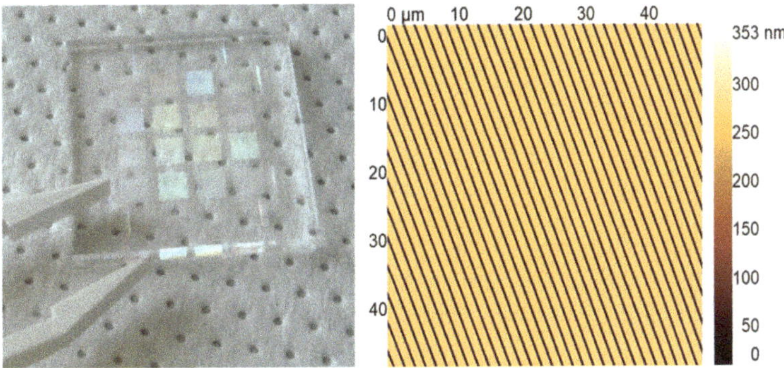

Figure 2. Left: photograph of the grating master. Right: AFM image of this master.

Again, OrmoComp was used for the final imprinting with the PDMS grating stamp. For the LTIL process, OrmoComp was diluted 1:2 with PGMEA (Propylene glycol monomethyl ether acetate) and then spin coated at 5,000 rpm on a silicon wafer. The solvent was then evaporated at 100 °C on a hotplate for 1 min. Prior to spin-coating, the wafer was treated with Profactor's HMNP-12 adhesion promoter [32]. The PDMS stamp was carefully brought into contact with the coated wafer and then peeled off again. Then the stamp was used for UV-NIL on top of the diffusor substrate. Pressure was applied using a self-built air pressure-based imprinting setup, which allows very homogeneous pressure distribution even on curved substrates [33]. UV-curing was performed again using the high-power UV-LED system at 365 nm.

The whole process sequence is schematically sketched in Figure 3, beginning with the fabrication of the diffusor substrate (step A (diffusor stamp fabrication) and step B (nanoimprinting of OrmoComp on PVC)). Step C is the fabrication of the PDMS stamp from the grating master, which is then used in the LTIL process (D-H). First, the material transfer from the donor wafer to the stamp for reversal NIL is accomplished (steps D and E). The coated stamp is then brought in contact with the prestructured substrate (F, G). By applying external pressure, the flexible PDMS stamp conforms to the topography of the underlying substrate (G), the OrmoComp is cured (G) and the stamp is removed (H), finalizing the LTIL process. The sample now contains the hierarchical structures.

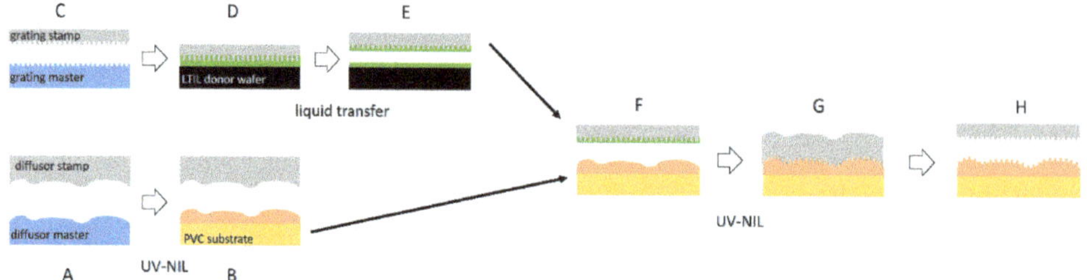

Figure 3. Process flow for master fabrication: First, a PDMS stamp from the diffusor master was fabricated (**A**), which is then used to imprint OrmoComp onto a PVC foil (**B**). Using a grating master, another PDMS stamp was made (**C**), which was used in an LTIL process (**D**,**E**) to print on the diffusor substrate (**F**–**H**). The flexible stamp conforms to the topography of the diffusor substrate as sketched in G. The results are the grating structures on top of the diffusor structures (**H**).

To avoid the trapping of air bubbles (especially in step G), two aspects are important. Firstly, the grating stamp has to be flexible enough to allow real conformal contact. In our case, the PDMS stamp was approximately 2 mm thick; for other substrate geometries thinner (i.e., more flexible) stamps will be necessary. Secondly, the contact geometry is critical. We bring the stamp into contact with one side of the substrate and then carefully lower the rest of the stamp in a lamination-type of process. Furthermore, nanoimprinting was performed in a vacuum [33]. Finally, the fact that PDMS is gas-permeable to some extent also helps in avoiding air bubbles [34,35].

The imprinting process was performed in such a way as to obtain four different areas on a single sample: the unstructured area, the diffusor structure alone, the grating structure alone, and an area with the hierarchical structure of grating on top of the diffusor. This was achieved by aligning the stamp to the substrate in such a way that the overlap between the diffraction grating pattern and the diffusor pattern was only achieved in a small area of the substrate.

From a sample that was prepared as described above, it is now possible to replicate a stamp for further nanoimprinting. A PDMS copy was prepared using conventional Sylgard 184 PDMS (1:10 mixing ratio, 23 h at 40 °C in a laboratory oven) as described above.

This stamp now contains the hierarchical structures and can be used in a conventional single step nanoimprint process. The imprint was again performed using OrmoComp as imprint material and a PVC foil as substrate. Figure 4 shows the processing sequence. In steps A–C the hierarchical structures sample is used as a master for working stamp fabrication. This working stamp is then used in a conventional Nanoimprint process (D–E). The advantage is that now the hierarchical structures can be replicated in a single nanoimprinting step.

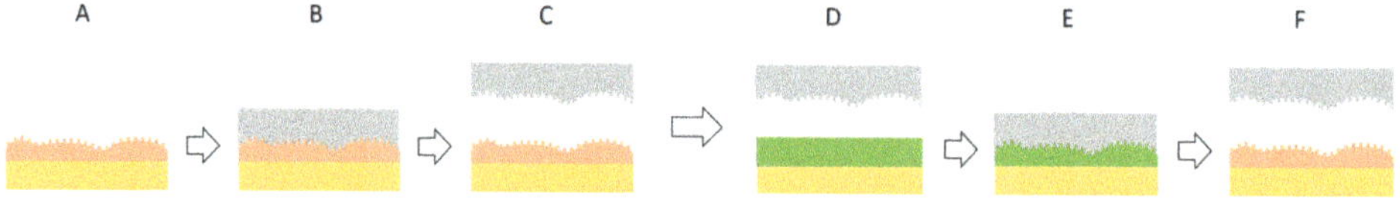

Figure 4. Process flow for stamp fabrication (**A**–**C**) and single step nanoimprinting of hierarchical structures (**D**–**F**).

3. Results

3.1. Hierarchical Stamp Master

Figures 4–8 show the result of these experiments. Figure 4 shows an optical micrograph image of the stamp master fabricated by combining conventional UV-NIL and LTIL. Both the diffusor structures and the grating structure can clearly be seen. AFM images from different positions of the same sample are shown in Figure 5. The grating structure follows the topography of the underlying diffusor structure in a conformal way and is present both in the valleys and on the hills. Looking at the line scan from one of the AFM images (Figure 6), it can be seen that the height of both types of structures is preserved.

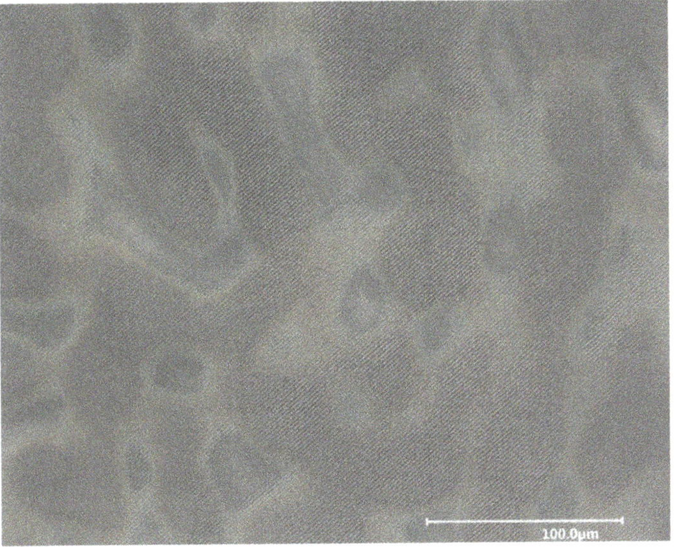

Figure 5. Optical micrograph of the grating on top of the diffusor structure. The lines and spaces follow the topography of the surface.

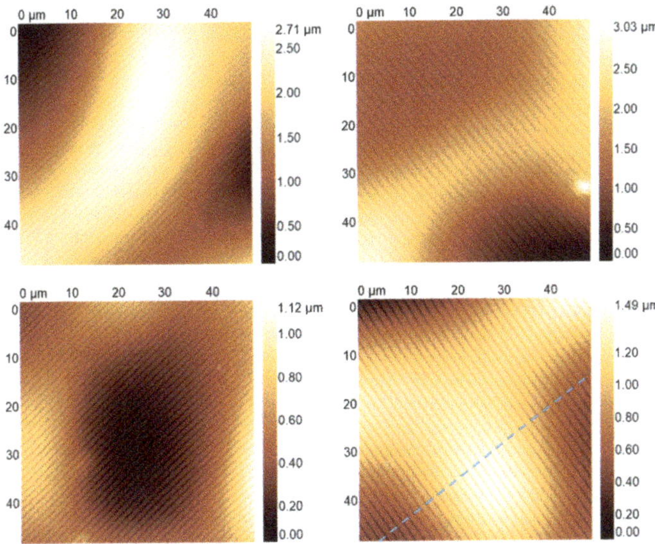

Figure 6. AFM images of the hierarchical structures master used for further imprinting. The left and right images have line and space dimensions of 300 nm/900 nm and 400 nm/1200 nm, respectively.

The dashed line in the lower right corner indicates the location of the extracted linescan shown in Figure 7.

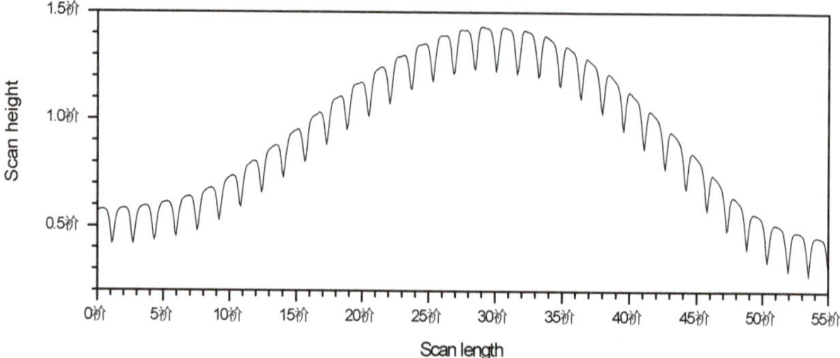

Figure 7. Line scan of the hierarchical structures master shown in Figures 5 and 6 above. The height of both types of structures is well preserved from the individual masters to the hierarchical structure.

Figure 8 compares the optical effects of the four different areas on the sample using a standard green laserpointer (λ = 532 nm). It can be seen that in the area with the hierarchical structures, both effects—the one of the diffusor and the one of the grating—are present.

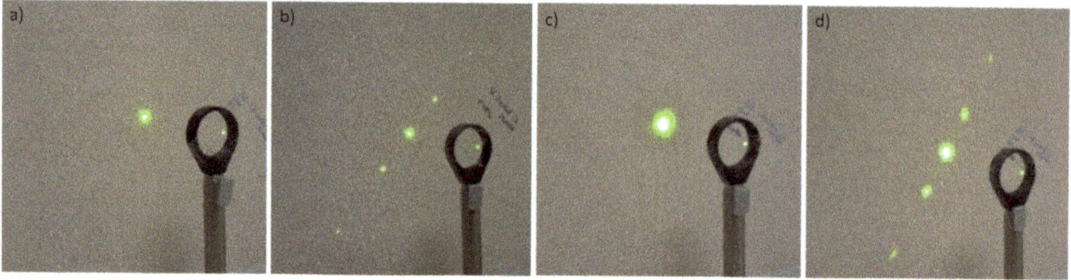

Figure 8. Photographs illustrating the optical effect of the differently stuctured areas by passing a green laser beam through the sample in 4 different areas: (**a**) unstructured, (**b**) only grating, (**c**) only diffusor and (**d**) grating on top of diffusor. The difference between b and d can be clearly seen.

3.2. Direct Hierarchical Nanoimprint

Figure 9 shows a compilation of four AFM images of the final sample, which was fabricated by a single imprint with a stamp copied from the hierarchical master sample. Due to the irregular structure of the diffusor, it was not possible to perform the measurements on the exact same spots as for the master again. However, our results show that the single step nanoimprint replication of hierarchical structures could be achieved with high fidelity. The line scan in Figure 10 shows the same characteristics as that produced by the flat grating master, the intermediate hierarchical structure master, and here on the final imprint. By evaluating the peak positions of the Fourier transformed images in Gwyddion [36], it was verified that the periodicities (1.60 μm or 1.20 μm, respectively) of the regions used for the line scans are identical on the grating master, the hierarchical structures master, and on the final imprint within the measurement error (+/- 60 nm, evaluated from 10 independent measurements using the 2D FFT function in Gwyddion). The volume shrinkage of the material, which is around 5%-7% according to the datasheet, does not play a significant role here, since the material is a thin layer fixed to a rigid, much thicker substrate. Shrinkage therefore will mainly happen in z-direction, since x- and y-directions are fixed. Due to the irregular nature of the diffusor structure, this effect could not be observed.

As far as the lifetime of the PDMS stamp is concerned, it strongly depends on the imprint resin used [34,37–40]. Furthermore, there are different types of PDMS that can be used, which also have different mechanical properties, e.g., h-PDMS [41,42] or X-PDMS [43].

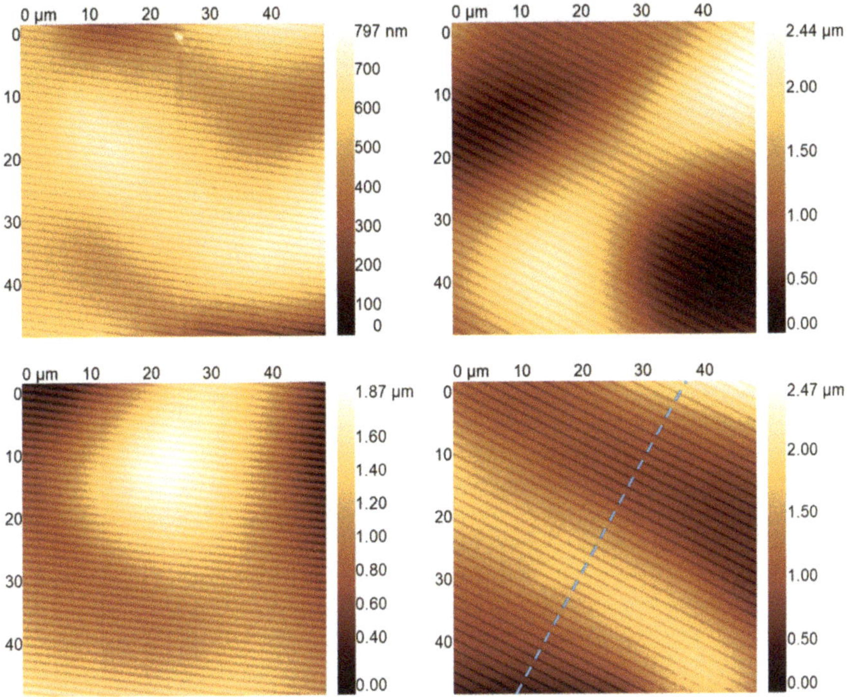

Figure 9. AFM image of the final imprint. The left and right images have line and space dimensions of 300 nm/900 nm and 400 nm/1200 nm, respectively. Comparing these with the results from Figure 5, it can be seen that the hierarchical structures could be replicated with high fidelity. The dashed line in the lower right corner indicates the location of the linescan shown in Figure 10.

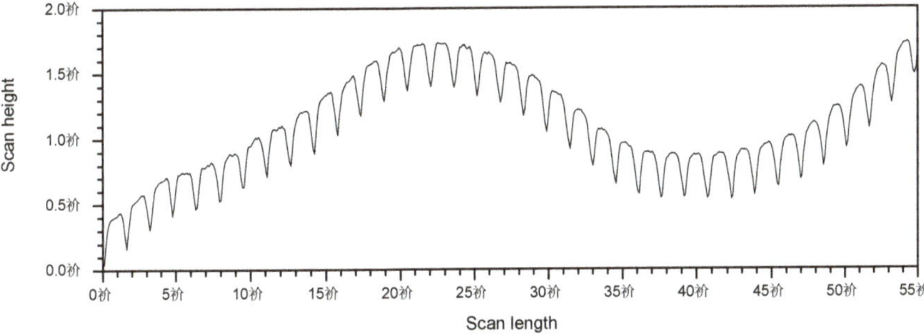

Figure 10. AFM linescan taken on the final imprint. The location is indicated by the dashed line in the lower right AFM image in Figure 9. Again, the height of both types of structures is well preserved from the individual masters to this final imprint.

4. Discussion

We showed that combining conventional NIL and LTIL is a suitable way to achieve multifunctional hierarchical surfaces. This has already been shown in other studies [26–28] and involves two separate nanoimprint steps and an unconventional nanoimprint method (LTIL). Using such a sample with a complex hierarchical geometry as a master for further stamp fabrication has several advantages in our opinion. To produce the final sample, the number of nanoimprint steps is reduced. Just one imprinting step is necessary instead of two, and consequently, the overall number of process steps is significantly reduced - typically by a factor of two - reducing cost of the overall process significantly. Material use is also optimized, since a spin-coating step can be omitted, resulting in a greener and more cost-effective process. Furthermore, potential adhesion problems between the two layers (diffusor layer and grating layer) can be avoided, although we have seen no evidence that this might be a problem with the materials we used in our experiments.

As an example, we demonstrated the combination of diffraction grating and diffusor structure. Such a sample was also successfully used as a master to fabricate a working stamp for nanoimprint lithography to replicate hierarchical structures in a single process step. The choice of structures for this work was based on the availability of masters and the possibility to easily distinguish between the optical effects of the two. In general, we are optimistic that the same process can also be used for combinations of other structures and also for other types of effects, ranging from antimicrobial to superhydrophobic. Moreover, applications like antireflective moth-eye structures on micro lenses should be feasible with this process. We have seen in different processes that PDMS stamps are capable of strong deformations, while still maintaining conformal contact. Examples are given in Appendix A.

Another interesting aspect of the use of such a master structure is that—assuming vertical sidewalls in the master and stamp for the grating—the hierarchical master should have negative sidewalls in many areas on one side of the grating structures. We have seen no evidence that the presence of such undercut features represents a problem for the soft PDMS stamp that we used. Both for larger and smaller structures, it has been shown that PDMS or other types of soft stamp materials are capable of replicating such features [44,45]. We will continue to work on combining different types of structures.

Furthermore, we will also work to apply this process to larger areas (DIN A4 and more), to be able to implement such stamps in our roll-to-plate nanoimprint process.

In conclusion, we presented a nanoimprint-based process for the creation and cost-efficient replication of hierarchical surfaces, demonstrated using a diffusor and a diffraction pattern. Using two subsequent nanoimprint steps, we realized a master structure that was further used to replicate a stamp for nanoimprinting. With this stamp we were able to create the structure that was initially created by two separate nanoimprinting steps using one single step.

Author Contributions: Conceptualization, M.M.M.; investigation, A.R.M, H.M.A., T.M., M.J.H.; writing—original draft preparation, A.R.M.; writing—review and editing, M.M.M.; project administration, M.M.M.; funding acquisition, M.M.M. All authors have read and agreed to the published version of the manuscript.

Funding: This research was funded by the ANIIPF project (bmvit – Austrian Ministry for Transport, Innovation and Technology) as well as by the rollerNIL project (FFG - Austrian Research Promotion Agency grant number 843639) and the NEAT project (FFG - Austrian Research Promotion Agency grant number 871438).

Conflicts of Interest: The authors declare no conflict of interest.

Appendix A

To further illustrate the deformation properties for PDMS stamps, some examples are given here. The UV-NIL process used here was slightly different. The imprint material OrmoComp was droplet–dispensed, and the nanoimprinting was performed using a vacuum–based process with a tool described elsewhere in detail [33]. The substrate was generated using glass spheres glued to a glass substrate. Gluing was accomplished using a layer of OrmoComp. Due to the strong deformation of the PDMS stamp, the stamp stretches on top of the structures, as illustrated in Figures A1–A3. Figure A2 shows an additional SEM image of the sample shown in Figure A1. Figure A3 shows an optical micrograph of an additional sample.

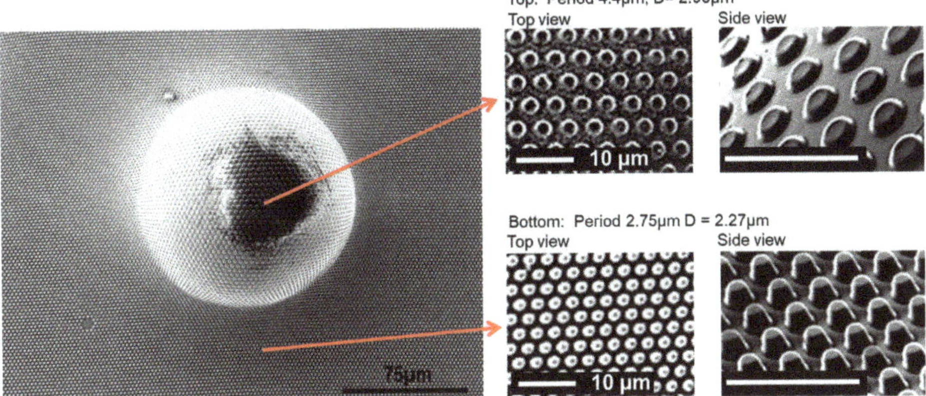

Figure A1. SEM images illustrating the deformation capabilities of PDMS stamps in nanoimprint lithography. The pattern on top of the sphere is significantly stretched. Consequently, the structures (pillars on the imprint, holes in the stamp) are also distorted. The period on the top is 160% of the period on the bottom, the diameter D of the features is also increased. The scale bars in the right images, which are cutouts from the left image and Figure A2, are 10 µm.

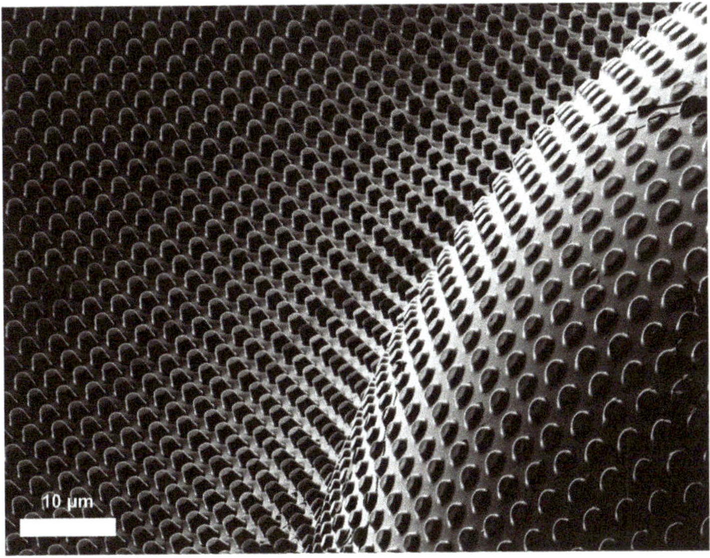

Figure A2. SEM image of microstructures nanoimprinted on top of a microsphere.

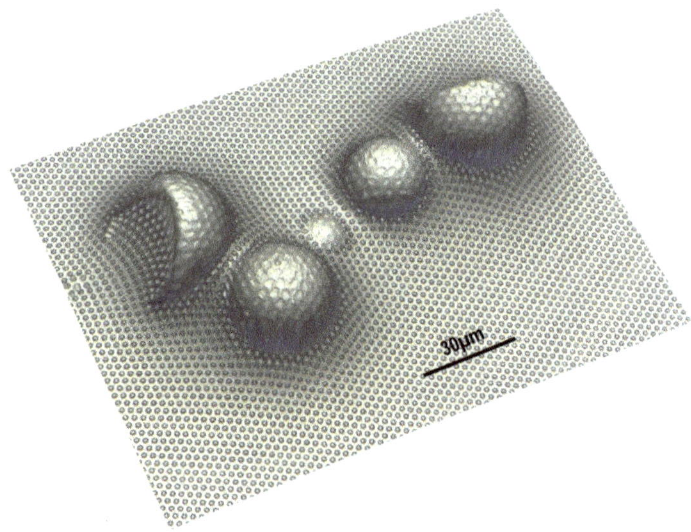

Figure A3. Optical micrograph (3D image stacking using a Keyence VHX-5000 digital microscope) of different types of microspheres over-imprinted with a PDMS stamp. In contrast to Figures A1 and A2 here, the stamp contained pillars instead of holes. It can clearly be seen that the pillars on the stamp are less stable under the imprinting pressure and collapse on top of the spheres, whereas the holes in the stamp remain more stable as can be seen in Figure A2.

References

1. Schift, H. Nanoimprint lithography: An old story in modern times? *A review. J. Vac. Sci. Technol. B Microelectron. Nanometer Struct.* **2008**, *26*, 458. [CrossRef]
2. Haisma, J. Mold-assisted nanolithography: A process for reliable pattern replication. *J. Vac. Sci. Technol. B Microelectron. Nanometer Struct.* **1996**, *14*, 4124. [CrossRef]
3. Chen, Y.-P.; Lee, C.-H.; Wang, L.A. Fabrication and characterization of multi-scale microlens arrays with anti-reflection and diffusion properties. *Nanotechnology* **2011**, *22*, 215303. [CrossRef] [PubMed]
4. Xie, S.; Wan, X.; Yang, B.; Zhang, W.; Wei, X.; Zhuang, S. Design and Fabrication of Wafer-Level Microlens Array with Moth-Eye Antireflective Nanostructures. *Nanomaterials* **2019**, *9*, 747. [CrossRef] [PubMed]
5. Reboud, V.; Obieta, I.; Bilbao, L.; Saez-Martinez, V.; Brun, M.; Laulagnet, F.; Landis, S. Imprinted hydrogels for tunable hemispherical microlenses. *Microelectron. Eng.* **2013**, *111*, 189–192. [CrossRef]
6. Glinsner, T.; Kreindl, G.; Kast, M. Nanoimprint Lithography: The technology makes its mark on CMOS image sensors and in the nano-world. *Opt. Photonik* **2010**, *5*, 42–45. [CrossRef]
7. Kreindl, G.; Glinsner, T.; Miller, R.; Treiblmayr, D.; Födisch, R. High accuracy UV-nanoimprint lithography step-and-repeat master stamp fabrication for wafer level camera application. *J. Vac. Sci. Technol. B Nanotechnol. Microelectron. Mater. Process. Meas. Phenom.* **2010**, *28*, C6M57–C6M62. [CrossRef]
8. Gale, M. Replication techniques for diffractive optical elements. *Microelectron. Eng.* **1997**, *34*, 321–339. [CrossRef]
9. Gale, M.; Rossi, M.; Rudmann, H.; Saarinen, J.; Schnieper, M. Replicated diffreactive optical elements in consumer products. In Proceedings of the Diffractive Optics and Micro-Optics, Rochester, NY, USA, 10–13 October 2004; OSA Publishing: Washington, DC, USA.
10. Osipov, V.; Doskolovich, L.L.; Bezus, E.A.; Drew, T.; Zhou, K.; Sawalha, K.; Swadener, G.; Wolffsohn, J.S.W. Application of nanoimprinting technique for fabrication of trifocal diffractive lens with sine-like radial profile. *J. Biomed. Opt.* **2015**, *20*, 025008. [CrossRef]
11. Schleunitz, A.; Schift, H. Fabrication of 3D nanoimprint stamps with continuous reliefs using dose-modulated electron beam lithography and thermal reflow. *J. Micromech. Microeng.* **2010**, *20*, 095002. [CrossRef]

12. Gilles, S.; Meier, M.; Prömpers, M.; van der Hart, A.; Kügeler, C.; Offenhäusser, A.; Mayer, D. UV nanoimprint lithography with rigid polymer molds. *Microelectron. Eng.* **2009**, *86*, 661–664. [CrossRef]
13. Muehlberger, M.; Boehm, M.; Bergmair, I.; Chouiki, M.; Schoeftner, R.; Kreindl, G.; Kast, M.; Treiblmayr, D.; Glinsner, T.; Miller, R.; et al. Nanoimprint lithography from CHARPAN Tool exposed master stamps with 12.5nmhp. *Microelectron. Eng.* **2011**, *88*, 2070–2073. [CrossRef]
14. Junarsa, I.; Nealey, P.F. Fabrication of masters for nanoimprint, step and flash, and soft lithography using hydrogen silsesquioxane and x-ray lithography. *J. Vac. Sci. Technol. B Microelectron. Nanometer Struct.* **2004**, *22*, 2685. [CrossRef]
15. Zhang, X.; Huang, R.; Liu, K.; Kumar, A.S.; Shan, X. Rotating-tool diamond turning of Fresnel lenses on a roller mold for manufacturing of functional optical film. *Precis. Eng.* **2018**, *51*, 445–457. [CrossRef]
16. Mühlberger, M.; Rohn, M.; Danzberger, J.; Sonntag, E.; Rank, A.; Schumm, L.; Kirchner, R.; Forsich, C.; Gorb, S.; Einwögerer, B.; et al. UV-NIL fabricated bio-inspired inlays for injection molding to influence the friction behavior of ceramic surfaces. *Microelectron. Eng.* **2015**, *141*, 140–144. [CrossRef]
17. Neinhuis, C.; Barthlott, W. Characterization and distribution of water-repellent, self-cleaning plant surfaces. *Ann. Bot.* **1997**, *79*, 667–677. [CrossRef]
18. Autumn, K.; Liang, Y.A.; Hsieh, S.T.; Zesch, W.; Chan, W.P.; Kenny, T.W.; Fearing, R.; Full, R.J. Adhesive force of a single gecko foot-hair. *Nature* **2000**, *405*, 681. [CrossRef]
19. Gao, H.; Wang, X.; Yao, H.; Gorb, S.; Arzt, E. Mechanics of hierarchical adhesion structures of geckos. *Mech. Mater.* **2005**, *37*, 275–285. [CrossRef]
20. Taillaert, D.; Bienstman, P.; Baets, R. Compact efficient broadband grating coupler for silicon-on-insulator waveguides. *Opt. Lett.* **2004**, *29*, 2749–2751. [CrossRef]
21. Daubinger, R. Hierarchical Nanostructures for Energy Devices. *Johns. Matthey Technol. Rev.* **2016**, *60*, 151–157. [CrossRef]
22. Li, W.-D.; Ding, F.; Hu, J.; Chou, S.Y. Three-dimensional cavity nanoantenna coupled plasmonic nanodots for ultrahigh and uniform surface-enhanced Raman scattering over large area. *Opt. Express* **2011**, *19*, 3925–3936. [CrossRef] [PubMed]
23. Micro Resist Technology GmbH | Hybrid Polymers. Available online: https://www.microresist.de/en/product/hybrid-polymers (accessed on 27 February 2020).
24. Huang, X.D.; Bao, L.-R.; Cheng, X.; Guo, L.J.; Pang, S.W.; Yee, A.F. Reversal imprinting by transferring polymer from mold to substrate. *J. Vac. Sci. Technol. B Microelectron. Nanometer Struct.* **2002**, *20*, 2872. [CrossRef]
25. Koo, N.; Wuk Kim, J.; Otto, M.; Moormann, C.; Kurz, H. Liquid transfer imprint lithography: A new route to residual layer thickness control. *J. Vac. Sci. Technol. B Microelectron. Nanometer Struct.* **2011**, *29*, 06FC12. [CrossRef]
26. Moormann, C.; Koo, N.; Kim, J.; Plachetka, U.; Schlachter, F.; Nowak, C. Liquid transfer nanoimprint replication on non-flat surfaces for optical applications. *Microelectron. Eng.* **2012**, *100*, 28–32. [CrossRef]
27. Lee, J.; Park, H.-H.; Choi, K.-B.; Kim, G.; Lim, H. Fabrication of hybrid structures using UV roll-typed liquid transfer imprint lithography for large areas. *Microelectron. Eng.* **2014**, *127*, 72–76. [CrossRef]
28. Uchida, T.; Yu, F.; Nihei, M.; Taniguchi, J. Fabrication of antireflection structures on the surface of optical lenses by using a liquid transfer imprint technique. *Microelectron. Eng.* **2016**, *153*, 43–47. [CrossRef]
29. OrmoComp®| Micro Resist Technology GmbH. Available online: http://www.microresist.de/en/products/hybrid-polymers/uv-imprint-uv-moulding/ormocomp%C2%AE (accessed on 9 October 2017).
30. Mühlberger, M.; Bergmair, I.; Klukowska, A.; Kolander, A.; Leichtfried, H.; Platzgummer, E.; Loeschner, H.; Ebm, C.; Grützner, G.; Schöftner, R. UV-NIL with working stamps made from Ormostamp. *Microelectron. Eng.* **2009**, *86*, 691–693. [CrossRef]
31. Beschichtungen BGL-GZ-83 | PROFACTOR. Available online: https://www.profactor.at/loesungen/beschichtungen/ (accessed on 27 February 2020).
32. Beschichtungen HMNP-12 | PROFACTOR. Available online: https://www.profactor.at/loesungen/beschichtungen/ (accessed on 27 February 2020).
33. Köpplmayr, T.; Häusler, L.; Bergmair, I.; Mühlberger, M. Nanoimprint Lithography on curved surfaces prepared by fused deposition modelling. *Surf. Topogr. Metrol. Prop.* **2015**, *3*, 024003. [CrossRef]
34. Toepke, M.W.; Beebe, D.J. PDMS absorption of small molecules and consequences in microfluidic applications. *Lab. Chip* **2006**, *6*, 1484. [CrossRef] [PubMed]

35. Lamberti, A.; Marasso, S.L.; Cocuzza, M. PDMS membranes with tunable gas permeability for microfluidic applications. *RSC Adv.* **2012**, *106*, 61415–61419. [CrossRef]
36. Nečas, D.; Klapetek, P. Gwyddion: An open-source software for SPM data analysis. *Open Phys.* **2012**, *10*, 181–188. [CrossRef]
37. Haslinger, M.J.; Verschuuren, M.A.; van Brakel, R.; Danzberger, J.; Bergmair, I.; Mühlberger, M. Stamp degradation for high volume UV enhanced substrate conformal imprint lithography (UV-SCIL). *Microelectron. Eng.* **2016**, *153*, 66–70. [CrossRef]
38. Haslinger, M.J.; Mitteramskogler, T.; Kopp, S.; Leichtfried, H.; Messerschmidt, M.; Thesen, M.W.; Mühlberger, M. Development of a Soft UV-NIL Step&Repeat and Lift-Off Process Chain for High Speed Metal Nanomesh Fabrication. *Nanotechnology* **2020**, (in press).
39. Schmitt, H.; Duempelmann, P.; Fader, R.; Rommel, M.; Bauer, A.J.; Frey, L.; Brehm, M.; Kraft, A. Life time evaluation of PDMS stamps for UV-enhanced substrate conformal imprint lithography. *Microelectron. Eng.* **2012**, *98*, 275–278. [CrossRef]
40. Tucher, N.; Höhn, O.; Hauser, H.; Müller, C.; Bläsi, B. Characterizing the degradation of PDMS stamps in nanoimprint lithography. *Microelectron. Eng.* **2017**, *180*, 40–44. [CrossRef]
41. Schmid, H.; Michel, B. Siloxane Polymers for High-Resolution, High-Accuracy Soft Lithography. *Macromolecules* **2000**, *33*, 3042–3049. [CrossRef]
42. Pei, L.; Balls, A.; Tippets, C.; Abbott, J.; Linford, M.R.; Hu, J.; Madan, A.; Allred, D.D.; Vanfleet, R.R.; Davis, R.C. Polymer molded templates for nanostructured amorphous silicon photovoltaics. *J. Vac. Sci. Technol. Vac. Surf. Films* **2011**, *29*, 021017. [CrossRef]
43. Verschuuren, M.A. Substrate Conformal Imprint Lithography for Nanophotonics. Ph.D. Thesis, Utrecht University, Utrecht, The Netherlands, 2010.
44. Möllenbeck, S.; Bogdanski, N.; Scheer, H.-C.; Zajadacz, J.; Zimmer, K. Moulding of arrowhead structures. *Microelectron. Eng.* **2009**, *86*, 608–610. [CrossRef]
45. Muehlberger, M. More than Just 2D: Nanoimprinting and Complex Geometries. In Proceedings of the 32nd International Microprocesses and Nanotechnology Conference (MNC 2019), Hiroshima, Japan, 28–31 October 2019.

MDPI
St. Alban-Anlage 66
4052 Basel
Switzerland
Tel. +41 61 683 77 34
Fax +41 61 302 89 18
www.mdpi.com

Coatings Editorial Office
E-mail: coatings@mdpi.com
www.mdpi.com/journal/coatings